Collectible
MICE

by Dr. Albert Eschen

Published by

Hobby House Press, Inc.
Grantsville, Maryland
www.hobbyhouse.com

Dedication

This book is for Flori, my wife of 50+ years. I am indebted to Flori who, through the years, has traveled thousands of miles with me and who has proven to be a better observer in detecting the elusive, rare mouse than I could ever be. The idea of writing a book about our collection came as a result of a gift my wife bought me, a word processor, and her telling me to get started. Therefore, if you like what I've written, I'll take the credit; if not, blame it on my wife. And, with love to Andrew, Burt, Caryn, Lowell and Meryl.

Needless to say, one of the primary functions of this book is to stimulate awareness, an interest in the study and collection of anything you enjoy gathering. I sincerely hope that anyone reading this book will get as much enjoyment from their collecting experience as I had in putting the book together.

Acknowledgments

A sincere thanks to the numerous dealers and manufacturers who so generously supplied me with information about their remarkable crafts. Full acknowledgment of all those who contributed to the development of this book would be impossible. I'm particularly grateful to: The Gorham Division of Textron, Inc., Providence, R.I.; Hummelwerk, Division of Goebel, Elmsford, N.Y.; Cybis Studio, Trenton, N.J.; President, Boehm Studio, Trenton, N.J.; Ray Blackman, V.P. Brielle Galleries, Brielle, N.J.; David Buten, Director, Buten Museum, Merion, PA.; Dan Judson, Heilgenthal Imports, Austin, Texas; Schmid Border Fine Arts, Randolph, MA.; Doulton & Co. Inc., Carlstadt, N.J.; Harmony Ball Co. Columbus, OH.

Mice Collectibles is an independent study by the author Dr. Albert Eschen and published by Hobby House Press, Inc. The research and publication of this book were not sponsored in anyway by the manufacturers of the items herein. Photographs of the featured items were from the collection of Dr. Eschen at the time the photographs were taken unless otherwise credited. The values given within this book are intended as value guides rather than arbitrarily set prices. The values quoted are as accurate as possible at the time of printing. In the case of errors, typographical, clerical, or otherwise, the author, publisher, nor manufacturers assume neither liability nor responsibility for any loss incurred by users of this book.

Mickey Mouse® and Minnie Mouse® are trademarks of Disney Enterprises, Inc. All rights reserved.

The metric conversions in this book have been rounded to the next whole number. To find an exact equivalent, multiply the number of inches by 2.54cm.

Additional copies of this book may be purchased at $24.95 (plus postage and handling) from
Hobby House Press, Inc.
1 Corporate Drive, Grantsville, MD 21536
1-800-554-1447
www.hobbyhouse.com
or from your favorite bookstore or dealer.

Printed in the United States of America

ISBN: 0-87588-654-X

Table of Contents

How I Became A Mouse Collector

When I was nine years old, I was presented with two white mice from my Aunt Carol and my Uncle Hank when I spent a summer vacation on their farm in New Brunswick, New Jersey. At that time, I dreamed of becoming a veterinarian, so I couldn't have received a better gift. I loved the mice immediately, and named them Fritz and Fred. Three weeks later, I was faced with the reality that Fred was actually Freda, and I was now the proud "papa" of six wiry white mice. When I returned to Brooklyn after the summer, my proud parenthood was quickly thwarted by my mother's ultimatum: either the mice go, or I go. At age nine, I had no choice; at age 55, however, I did, and my wife and I started our own mouse collection.

Our collection evolved out of our appreciation and pleasure that we derived from Edward Marshall Boehm porcelain sculptures. A friend of mine, an established Boehm dealer in Atlantic City, New Jersey, introduced us to the Boehm experience. When Flori and I discovered a Boehm figure that we both admired, we added it to our collection. In January 1976, my wife spotted a small porcelain mouse in a shop window on Worth Avenue, in Palm Beach, Florida. Recognizing it to be a mint, porcelain mouse made by Boehm in 1960, she purchased it for $450. To our delight, we later learned from a newspaper article that this figurine was valued at $2,500. Later that same year, while I was enjoying the magnificent glass-engraving exhibition at the Steuben glass gallery in New York City, I noticed a gold mouse sitting on a gleaming bright wedge of crystal cheese. This beauty became the second mouse in our collection.

Generally speaking, there are three kinds of mice.

One kind lives in the real world—the furry little cheese-eaters that scurry around. The second lives in the imagination, found in books, cartoons, and songs. The third fills the shelves of our curio cabinets. Once we got hooked on mice, we found that we were on our own when it came to knowing what to look for. As mouse collectors, we were not as fortunate as "cat collectors" or even "pig collectors" who could turn to a plethora of literature for guidance.

For the past 25 years we have traveled throughout the United States and Europe, constantly adding to our mouse collection. Along the way we have met many mouse collectors at flea markets, garage sales, and antique shows. One avid collector in South Florida has his swimming pool shaped like Mickey Mouse's head. A married couple I met who are also collectors put vanity plates on their cars reading "MOUSE #1" and "MOUSE #2". Today, our collection numbers over 1200 different mice—all special to us—and is still growing. It is an eclectic collection of ivory, brass, porcelain, glass, diamond, gold, wood, and pewter figurines. Several figurines are over 150 years old; others are as new as last month. Some cost over $3500, others as little as $20.

This book is a compilation of our 35 years of research and collecting. It is written for the novice as well as the seasoned collector. As you leaf through this book, please remember that the values stated are based on my opinion, and are premised on the particular item being in perfect condition. Imperfections, chips, paint scratches, dents, unintended discoloration and repairs will certainly devalue a piece. Prices paid at auction and to dealers will vary greatly and, ultimately, the sales price is determined by the strength of demand.

Boehm Field Mouse
Letter shown with price and appraisal value of Boehm porcelain figure. Dated January 9, 1975.

Personal Preferences In Building A Collection

No matter what you may choose to collect, it is up to you to determine the depth and breadth of your collection. Remember, you're in charge here. Much of your decision will be based on your lifestyle and, of course, the amount of money you may choose to spend. However, some of the most interesting collections that I have seen have been built on a shoestring budget. Lifestyle often involves how much time one has to spend on the actual search—the act of collecting. Before I retired in 1982 (I was an Optometrist), our collection totaled approximately 900 pieces. Between then and now, our collection has increased by at least 300 new pieces. But, no matter how much or how little time you may have to collect, I guarantee that you will enjoy every minute of it as long as you view it as an adventure. Your travels, for business or pleasure, will take on a new meaning as you will be on the lookout for the "perfect" addition to your collection.

A friend of mine, who is a marriage counselor, once told me that she recommends to all couples that she sees that they start a collection of whatever items may appeal to them. "It brings people closer together," she insists. I agree. Flori and I can spend hours reminiscing about the wonderful times we've had over the years just by looking at striking sculptures. "Remember this one?" she might ask. Of course I do. "It's the Vienna Bronze optometry examination room figurines I gave you for our 35th anniversary." (This

collection has extra, obvious significance to me because of my profession.) Then I point to another of our little prizes. "Remember when I was at that college reunion in Chicago, and you came back to the hotel all charged up because you found that quaint little out-of-the way antique shop?"

What does it take to be a collector? You have to be curious. You have to like the odd and the unusual. You have to enjoy the "hunt." That requires patience and diligence, as well as a sense of adventure. As your collection grows, you will become more discriminating, and it is the ability to discriminate which is the collector's most important asset.

Before you begin your collection, you may need to ask yourself whether you are collecting for the fun of it, or for financial gain. Will you be buying objects because you want to keep them for your own personal pleasure, or because you want to turn around (at some point) and try to make a profit? Whether you decide to collect for pleasure or for investment, I always recommend buying quality, mint pieces, whenever possible. When people hear me say this, they often respond with, "That's fine for you to say but I don't have the money." My response? "Sure you do." If you have a choice between buying one piece in good condition for $50, and two in inferior condition for $25 (or even $20) each, I suggest that you pursue the quality piece. Quality doesn't mean that the item has to cost hundreds of dollars. Similarly, if you have

Pied Piper of Hamelin
8³/₈in (21.27cm)
England
Marks/Description: Royal Doulton stamp;
lion over crown incised; handmade.
Ca. 1993
Value: $200

a choice between one original piece for $100, or two reproductions costing $50 a piece, I would urge you to purchase the original. In my years of collecting, I have never seen an imitation—although in perfect condition—worth as much as an original.

Many fine pieces have practically disappeared, but occasionally unique mice can still be found in the marketplace. It is virtually impossible to collect, classify, or photograph every quality figure that has been manufactured; therefore, I have set my sights on finding the most desirable, and sought-after collectibles. The beginner mouse collector can gain knowledge with the different manufacturers' styles and types by becoming familiar with some of the pictures in this book from our private collection. Pricing these figures is very difficult. Colors, crudeness and imperfections will definitely lower value. Mint pieces will, in the vast majority of cases, be more expensive. It is not the purpose of this book to set prices, but to let you know what I find in the marketplace. Most of the mice in this collection have initials, inscribed names, trademarks, or surface marks. These distinguishing characteristics must be considered when purchasing an expensive item. Beauty, design, color, and rarity determine the price and affect market value. Several of the mice pictured in this book can still be found in the marketplace.

In selecting pictures for this book I have tried to choose those figures that show different design features, colors, textures and–very important to me–humor. Some collectors have been mainly concerned with the ability to assemble large numbers of items, regardless of their size, shape, value, or quality. To me, this is "accumulating"; it is not collecting. (Later on I will identify by photograph and describe some of the more interesting pieces that we have been able to collect.) Assembling a large collection of mice depends, to a great extent, on your enthusiasm and watchful eye. A dealer with integrity, with whom you can establish a trusting relationship may keep you informed and let you know when he has a good "find" for you.

The discovery of a rare piece may be a happy moment following a hope or desire that you have held for many years. In my opinion, this level of satisfaction will never be understood by one who buys for investment purposes only. If you are purchasing an individual item or a complete collection with the intention of selling it (and this goes for any antique), my advice would be to know, in advance, to whom you are going to sell the object before you buy it. Obviously, be sure that you buy it at a price, which would allow for a reasonable profit.

You may want to limit your collection to a single medium, e.g., brass, gold, silver, porcelain, or bronze. This will make the hunt more limiting, but much more challenging. From my experience, there is enough in the marketplace to build a solid and rewarding collection based in any medium you may choose.

How "pure" do you want your collection to be? A friend of mine collects pig figurines, but refuses to allow any boars to get through the door. When you collect mice, you may find yourself faced with the "rat" question. Occasionally, figures have been designed to be mice, but the artist takes "poetic license" by adding a longer tail. Would you then consider it to be a "rat" and pass it by? That's a decision that only you can make. Although every purchase we made was with the thought that we were adding another mouse to our collection, we learned—a bit too late—that some rats had actually slipped in—usually following behind the Pied Piper of Hamelin.

We had always believed that the Pied Piper was leading mice out of town. Consequently, every time we found an interesting piece we purchased it. Over the years we purchased a total of six beautiful pieces before we learned that the Pied Piper was the leader of rats. For that reason, and in the interest of all mice collectors who will read this book, we decided to include some of these extraordinary pieces in this book. By the way, in defense of our purchases, D.G. Rice, in his 1989 book, *Encyclopedia of English Porcelain Animals of The Nineteenth Century Antique Collector's Club, England*, reports that many porcelain mice were well-known in the nineteenth century, and that it was most difficult, even at that time, to distinguish between a porcelain mouse and a

Vienna Bronze Orchestra
Average Height 2³/₁₆in (5.56cm)
Austria
Fritz Bermann Company
Marks/Description: "FBW" in circle; fourteen
musicians with instruments, two singers,
music holder, leader and bandstand.
Ca. 1930
Value: $4500

porcelain rat. (He also observed that there are many different types of mice, and that mice were often kept as pets.)

Every collector is sure to specialize. We had the good fortune to purchase a complete orchestra of Vienna Bronze figurines from a gentleman who collects only animal musicians. Collect those pieces that appeal to you. Act on your own good taste and instincts. Get involved in some facet of collecting, and share your experience with other collectors.

Some dealers have tried to sell us Mickey Mouse pieces. We, however, shy away from Mickey Mouse collectibles. In my view, there are so many Mickey Mouse collectors, so many clubs and societies, it would be futile to attempt to list them all in this book. Many of these clubs have their own magazines and special auctions. Some dealers absolutely refuse to have "any" Mickeys claiming that they are just "too commercial." That's the way we feel about it, too. When I walk through a shopping mall and can purchase at least one hundred different Mickey items (e.g., toys, books, clothing, t-shirts, cards) in less than one hour, I don't call that collecting. I have included in this book photos of a dozen or so Mickeys that my wife and I picked up this past week at a local flea market in about two hours, at a cost of less than $25. Now you can understand why we call this accumulating, not collecting.

On the other hand, I don't mind an antique Mickey Mouse piece (or a contemporary piece if it was made by a recognized artist). In that regard, I have included in this book photos of a pair of Mickey and Minnie shovels that I purchased several years ago from a dealer who advised that they are souvenirs from the New York World's Fair in 1939. I have also included pictures from another exceptional Disney group. This Mickey Mouse Vienna Bronze orchestra, I understand, was manufactured in the 1920s.

One couple we occasionally meet at antique shows told us that their kitchen is decorated with amusing mice. They have odd, comically shaped salt and pepper shakers, teapots, coffee pots, cookie jars, cheese boards, and other humorous utensil mice. Even their kitchen curtains have a Mickey Mouse pattern. And, a lady that my wife is acquainted with, collects and wears mostly mouse jewelry—and I don't mean costume jewelry or "toy stuff."

And then there's an exterminator around town who drives a bright yellow Volkswagen outfitted to look like a mouse. It has large, round, three-foot black ears attached to the roof, black whiskers painted on the hood, and a conspicuous black rope tail attached to its rear bumper.

The lesson is simple: Get involved in some aspect of collecting, and share your experience with other collectors!

Left:
Mickey and Minnie Mouse Shovels
4¹/₈in (10.48cm)
Marks/Description: none
Ca. 1935
Value: $35

Below:
Vienna Bronze Mickey Mouse Band
Austria
Fritz Bermann Co.
Marks/Description: "FBW" in circle;
fourteen-piece orchestra complete
with instruments on mouse-shaped
wood cheese board
Ca. 1935
Value: $4,500

How To Start A Mouse Collection

Some antique dealers recognize us and try to double the price of their mouse items; such is the peril of being known as serious mouse collectors. And, because there were no reference guides (until this book), it was often difficult to look at a mouse figurine and have significant bargaining power, especially if it was unknown what the market would bear. Of course, that's why it is so important to become familiar with the different manufacturers, the different materials, and other factors that affect value in a mouse collection.

Although we consider our collection to have some of the rarest mice in the world, many of the pieces were found in thrift stores, flea markets, garage sales and antique shows. Our collection does not include greeting cards, paper napkins, origami, or stationary decorated with humorous mouse sayings or drawings. It doesn't include mice in traps or mice being harmed in any way. In other words, we don't collect any cats or owls enjoying a mouse barbecue. We look for mice with pleasant—sometimes even comical—expressions on their faces.

By now you realize that one purpose of this book is to acquaint the novice collector with some of the unique items that are available. I'm sure the advanced collector will enjoy his or her collection even more by comparing it to my photographs.

Remember to examine your purchase carefully to determine its genuineness; insist that the seller provide a certificate, or other written document, attesting to its authenticity. An amateur will likely make errors since little knowledge may lead to hasty judgement. It is not possible to set up a standard of rules for recognition of the imitation, but there are usual signs that are gradually learned so that a collector can be aware of "improved" pieces. It is only through personal experience that a collector may learn whether a mouse is of some value.

Some dealers may attempt to deceive and elevate prices. Similarly, one problem that my wife and I run into occasionally is the antique dealer who recognizes us and knows that we collect these little critters. Helpful dealers will put aside what they think we would be interested in at a fair price. There are a few dealers who will try to offer some items to us at astronomical prices; while we admire these pieces, the unreasonable price commands that we do so only at a distance. We have traveled through about forty states and have vacationed in about fifteen countries, and have yet to come upon a location that hasn't been investigated by professional antique dealer. Fortunately for us we find that even a professional misses an occasional piece of art.

Some simple advice, if you want some pretty items and want them at a lower price, you're going to have to make an effort to find them. Get out and start looking for them. It's a great feeling and much more fun to locate a rare item than to pay a high retail price for it. I need not tell you how much satisfaction and personal triumph there is when you locate a "good" find at a low price. Three years ago while I was browsing in a second-hand store I noticed a bronze mouse statue. I made an offer and the owner gave me this "dust collector" for $38. The fair market price for this piece today is about $140.

Purchase only mint pieces, especially if you have any doubts. By recognizing the originals you can reduce or avoid the discouragement resulting from later discovering that you purchased copies or imitations. Seldom are collectible mice alike, yet by experience you will learn to recognize the unmistakable styles and hallmarks of a particular maker. You will learn those symbols that make for distinctive and interesting art—some rare, some modern. An old Vienna bronze or old porcelain statue in mint condition can be worth many times the cost of a gold one.

On some mice collectibles no date, trademark, or manufacturer may be found. To my knowledge, no other book on mice collecting has ever been written, let alone published, so the only way that I know how to identify a particular piece is by differentiation. Carefully compare size, shape, color, material, and any other specific identifying characteristics with other pieces in your collection. These will, perhaps, give you a clue as to the artist and the country of origin. The surest way for a collector to identify an object is by the manufacturer's trademark. Since not all models have marks, the photographs that I've taken will make identification of originals much easier.

Two Mice on Sealing Wax Signet
England
Marks/Description: malachite stone; initials JC; Lionel Flowers and letter "E" Vickery Regent St.; engraving of greyhound on bottom of base.
Ca. 1890
Value: $650

As you embark on making your own collection, you should also be aware of some important concepts used by dealers:

- *Limited edition*: An edition of a contemporary product. The manufacturer announces the total number of the edition at the time of production.
  Non-limited edition: An edition in which no total number is announced. It can be discontinued at any time or the manufacturer may produce more at his own discretion.
- *Open edition*: Although a limited edition may be all sold out, the manufacturer may still be producing them.
- *Closed edition*: Limited and non-limited editions can fall in this category. A limited edition is closed when the total number of objects has been made. A non-limited edition is closed when production is stopped.
- *Secondary market*: The market in which an object enters <u>after</u> its first sale from the dealer to the collector. The first collector may sell it to another collector or dealer. The price is then determined by supply and demand. Secondary markets frequently command considerably higher prices than an original issue price.
- *Mint condition:* Completely without flaw.
- *Makers mark*: These are the marks fired onto the bottom of a piece of porcelain or the impression left by the die when bronze, pewter, and other metallic surfaces are manufactured. By understanding these marks it is often possible to trace the history of the piece and the manufacturer. Note: Because a particular company may have used different marks (or variations of a mark) over the years, it is not possible in this book to describe all of them. An excellent reference book that depicts manufacturers' marks is the *Dictionary of Marks - Pottery and Porcelain* by Ralph and Terry Kovel. I recommend that the collector consult that treatise (or other such compilations) for further information on the marks that may have been used.

Japanese Silver Three-Piece Tea Set
Japan
Marks/Description: Excellent quality Japanese tea-set humorously decorated with mice stealing money (Koban) with woven rope and tassel handles on each of the three pieces, and mouse finials on the teapot and covered sugar bowl. Marked in Japanese (Jun-Gin) pure silver. Imported into England early 20th century and hallmarked

Ca. 1895
Value: $4500

It is rather difficult to say what is worthwhile and what isn't. What you might like to look at, or hold, doesn't necessarily mean that others will appreciate it. Don't buy just to buy. Collect, don't accumulate. I have always found auctions to be fun places, enjoyable experiences. I guess the satisfaction of purchasing something at a low bid is exciting for everyone. However don't get sucked in. If you feel the item is overbid forget it. There will be other days and other times. As I mentioned before, purchase good quality and workmanship. The truth of the matter is that collectors often pay dealers considerably more than they would pay a private collector where the object has been cared for and loved for many years.

When making a purchase use good judgement as to the reliability of the seller and the genuineness of the article. Look for hallmarks, initials and surface marks especially when considering the purchase of an expensive mouse. Beauty, design, color, rarity, and mint condition determine the price and true value. Insist on a certificate of authenticity from the person from whom you purchase the object.

I've researched manufacturers and artists whose work I've been able to identify. These are the pieces with marks, country of origin, and some with artists' names. Unfortunately, many mice have no obvious marks or symbols, and in several cases it was impossible to make positive identification. In conducting research for this book, I have approached and spoken with many dealers and collectors. By comparing shapes, sizes, colors and texture I've presented what I believe to be fair recognition of these exceptional pieces. Incidentally, just because a mouse has no identifying marks doesn't mean that it's not top quality. Many are first-class, but, with identifying information, they would be worth considerably more.

ELEVEN COLLECTING TIPS

In summary, I believe that the following tips will enhance your collection and the whole experience of collecting mouse figurines:

1. Acquire as much information as you can before you start collecting.

2. Never stop learning about your collection.

3. Develop a relationship with at least one reputable dealer.

4. Try to continually upgrade you collection. If necessary, sell or trade your less valuable pieces so you can be in a better position to buy more valuable pieces.

5. Catalog your items and take photographs of each piece. Your filing system should include the price, place, and date of purchase, and background information on the piece. As you learn more about the piece, especially its (increasing) value, update your catalog.

6. Don't buy an item only because it's old or appears to be a bargain. Buy it because you like it.

7. Trust your own good judgement and taste as well as your knowledge.

8. Maintain the integrity of your collection. Just because a friend buys you an item, it doesn't necessarily belong in your collection. In other words, don't compromise your collection because you're trying to be polite.

9. Don't become discouraged. Learn from your mistakes.

10. Find unique ways to display your collection.

11. Talk to your insurance broker or agent to make sure that your collection is covered by a policy that affords coverage against loss or damage.

The Mouse In Literature and Folklore

As our collection became larger and larger, and as we did more research into the pieces, we became more and more interested in how the lowly mouse has been described over the centuries. There are many species of mice. The *Harvest Mouse* is the smallest and also the most lively in color. It measures about 2¼ inches long with a 2¼ inch tail. This mouse is usually of a gold-orange-brown color. In warm weather this fellow lives out on the farm, in and around the corn and wheat fields, and in the tall grass. These interesting little mice are clean, adapt quickly, and soon get very accustomed to captivity.

The *Wood Mouse* has also been called the Field Mouse, and is a little larger than the Harvest Mouse. This fellow lives in fields, gardens, in the woods, and occasionally at the beach. He's usually a night creature and excels in climbing to obtain most of his food. The Wood Mouse breeds several times during the year, and usually has four to six young at a time.

Then, there's my favorite: the *Common* or *House Mouse*. Smaller than the Wood Mouse, the House Mouse usually has a dull uniform color (brownish-gray), with a short tail and ears. (As a retired optometrist, I am particularly partial to the Common Mouse because of its "beady" eyes.) The House Mouse likely had its origin in Asia. It usually makes a home for itself in houses and occasionally will live it up even before the house is completed. It is less nocturnal than the Wood Mouse and, on occasion, will look for food in the daylight. There are, of course, many other species of mice.

In the beginning, the mouse got a bad rap. Some legends suggest that the devil created himself as a mouse because he was not pleased with Noah. Mythological stories propose that the devil was hoping that all God's creatures would drown so he set out to sabotage the Ark by chewing its timbers. Another myth (from Germany) reports that mice fell to earth from clouds during a storm. Yet another myth attributes the advent of mice to witches who made them from a piece of cloth. Later, these mice ran away. Did you know that in Egypt, the mongoose has been called "Pharaoh's cat?" It is better known as "Pharaoh's Mouse" for its cleverness in catching mice. (If you have a special interest in myths, you may want to peruse *Funk & Wagnall's Standard Dictionary of Folklore Mythology and Legend*.)

Other mouse legends recite the tale (or perhaps, more appropriately, the tail) about the girl who threw little balls of dirt over her shoulder into a field. When she returned, the field was full of mice. According to Jewish legend, the mouse is considered a bad omen if it gnaws on the clothing of a sleeping person; it was seen as a sign of death. Similarly in Jewish folklore, anything gnawed or nibbled on by a mouse is never eaten by humans: this food can cause forgetfulness or a sore throat. In Greek legend, some believed that the mouse had an evil eye. The mouse was pictured on their coins as a protection against evil. Funk and Wagnall's also reports that white mice in some parts of Bohemia and Germany were fed and kept alive in people's homes because they are believed to bring good luck.

Other legends relay mice as ministers of vengeance. These were the mice who plagued the Philistines who had taken the Ark of the Covenant from the camp of the Israelites. In France and Northern Germany it is said that the appearance of great numbers of mice is an omen of war.

Although presently, mice are used in medical experiments, in previous ages, they were used as cures. In England, for example, mouse pie or fried

Cinderella's Coach
Mexico
Marks/Description: sterling; 925; Taxco; letters "BCS";
chariot driven by two mice and two coachmen
Ca. 1935
Value: $20

mice was prescribed as a cure for bed-wetting. Small-pox, whooping cough, and measles were also cured with cooked mice. Pliny recommended mouse ashes mixed with honey for an earache and as a guarantee of sweet breath if used as a mouthwash.

How can we forget the mice in the Cinderella story by Charles Perrault. Remember the part where the fairy godmother with a touch of her magic wand, turns the mice into six prancing horses?

Then there's the two newly written fables by Hazel Shertzer. One is an African folktale where the mouse finds a lion trapped in a net. The mouse remembered that the lion once spared his life and so he gnaws the ropes and frees the lion. Moral: It's good to help and be kind to each other. The other story tells of the cat that ate the mouse; the mouse, of course, erroneously thought the cat was friendly and good natured. Moral: The look of something can mislead, and the look in a face not always reflects the mind.

Jeremy Mouse by Patricia M. Scarry has twenty charming mouse stories and drawings for pre-school and grade school children.

By now you realize that most of the material in this book is not original. I have spent a considerable amount of time in libraries and bookstores searching. I have tried to sort out as much interesting and curious information about mice as I could, and I assure you the statistics I have supplied are reliable.

Of course, the little mouse has worked its way into our daily expressions. Who hasn't heard the following expressions: "Quiet as a mouse," "mousey colored hair," "poor as a church mouse," "quick as a mouse" and "timid as a mouse." A dark swollen bruise under the eye is sometimes called a mouse. And a "mouse," years ago, was a common term of endearment, like birdie, duckie, or lamb. (Then of course, there is the ubiquitous mouse found in close proximity to every computer in the world.)

Bartlett's Familiar Quotations
lists the following famous quotations involving the mouse:

"She watches him, as a cat watches a mouse." (323.19)

"Pussycat, pussycat where have you been—I frightened a little mouse under the chair." (932.8)

"I am the magical mouse—I don't eat cheese—I eat sunsets—and the tops of trees." (879.14)

"Life goes on forever like a gnawing mouse." (822.16)

"'Twas the night before Christmas, when all through the house not a creature was stirring—not even a mouse." (446.3)

"Hickory, dickory dock, the mouse ran up the clock." (930.4)

And, the following quotations can be found in
The Oxford Dictionary of Quotations:

"Good my mouse of virtue, answer me." (*Twelfth Night*) (88.7)

"Not a mouse shall disturb this hallowed house." (*Much Ado About Nothing*) (471.22)

"Not a mouse stirring." (*Hamlett*) (312.23)

I now turn to the more serious information,
vital to anyone considering collecting mouse figurines.

What To Look For In A Mouse Collectible

THE GLASS MOUSE

Before 1800, the rough spot on the bottom of a piece of finished glass was called the pontil mark, which was caused by the rod that held the glass while it was hot and being worked on. In the late 1800s it was ground down and smoothed over. To identify quality glass pieces, you need to look for:

- Pontil mark: this will give you approximate date of manufacture.
- Wear and tear: chips and cracks reduce value of any kind of glass.
- Color: clear glass—transparent glass to which no color has been added—may be colorless, or on cheaper batches of glass, may show a green, yellow or brown tint which comes from the elements.
- Weight: flint glass is heavier, brilliant and colorless due to special ingredients.
- Sound: listen for the clear rich sound when you flick it. (Experts caution not to do this for the obvious reason.)

The glass mice in our collection were made by the following manufacturers:

BACCARAT

Baccarat glass was first made with a leaded composition in France in 1765. It was not until 1820, however, that this type of glass began to be respected and admired. Baccarat produces mostly crystal glass, however various colored glass is also made. The Baccarat company is still working near Paris and primarily makes crystal figures, glassware, and paperweights. Each piece has the signature Baccarat signed in script. The only true way a collector can guard against imitations and counterfeits (short of asking Baccarat itself to authenticate it or possessing a certificate of authenticity) is by frequent handling and the careful examination of old authentic pieces.

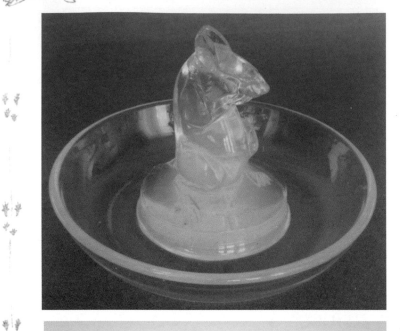

Lalique Opalescent Card Tray
3³/₄in (9.53cm)
France
Marks/Description: R. Lalique, France
Ca. 1935
Value: $850

LALIQUE

René Lalique (1860-1945) started experimenting with glass around 1897. All types of glassware has been manufactured by this family-owned company, but the frosted finish style that he placed on many of his creations became associated with his name. The company mainly produces colorless glass, but very fine opalescent pieces are also manufactured. The mark "R Lalique" was changed to "Lalique" after he died in 1945. His granddaughter was also its top designer. Lalique glass, besides being molded, is pressed and also engraved in Art Nouveau and Art Deco styles. Design numbers are occasionally included. Hand-blown glass is signed in script and after 1926, the word "France" was also included.

Steuben Crystal Cheese and 18K Gold Mouse
4in (10.16cm)
New York, NY
Marks/Description: diamond point signature on base "Steuben";18K gold mouse sitting on wedge of sliced and polished crystal cheese. Designed by James Houston. James Houston was born in Canada. He studied painting and graphic arts in Canada, Paris, and Tokyo. He joined the staff of Steuben Glass in 1962.
Ca. 1975
Value: $1,150

STEUBEN

Steuben Glass has been in Corning, New York, for about 90 years. It is an American company that makes all its glass in Corning. It is named after Steuben County where Corning is located. Steuben Glass was founded in 1903 by Frederick Carder. Steuben produces only clear, colorless, highly reflective crystal. Unsurpassed quality of material, design, and craftsmanship continues to be its minimum standard. At Steuben, glass is produced virtually free of bubbles, discoloration and other imperfections. Steuben, in my opinion, is unequaled for brilliance. The Steuben name, signed by hand with a diamond point on each piece, confirms the exacting standards by which each design is made. Steuben's designers follow their own artistic leanings. They make only a limited number of pieces each year striving for quality rather than quantity. You will find no "factory seconds" at Steuben. Having personally studied glass properties for the past forty years, in my opinion, Steuben is the finest crystal manufactured today. I regard their crystal so highly that we have six of their mice in our collection.

SABINO

Marius Ernest Sabino, who founded the Sabino company, started making glass in Paris, France in the late 1920s. The firm closed its operation during World War II but started up again in the 1960s with the manufacture of small opalescent animals. Each sculpture is marked "Sabino France" and his signature. New pieces are recognized because of a slight difference in color.

Opalescent Mouse
Sabino, France
Marks/Description: Sabino, France. Sabino glass was made in the 1920's in Paris, France. Production halted during WWII, but resumed about 1960. New pieces are readily recognized because of a slight difference in color.
Ca. 1975
Value: $70

DAUM

The Daum Glass Company was established in 1875 in Nancy, France. Daum manufactured cased glass, etched glass, and frosted types. The Daum signature in various forms was usually signed on the base. The mark became "Daum Nancy" with the Cross of Lorraine on the base around 1910. "France" was added after 1919.

Multi-Color Glass Mouse
5in (12.7cm)
France
Marks/Description: Daum Nancy. Auguste and his brother Antonin, Cameo glass artists, signed each piece of work after 1875.
Ca. 1993
Value: $195

SWAROVSKI

One of the world's largest glass cutting factories is situated in Wattens, Austria, a small Tyrolean village. The company has been producing full lead crystal since 1889. The remarkable precision of the facets enables each piece to refract sunlight perfectly. Swarovski Silver Crystal is the culmination of unit-purist raw material plus perfection in cutting and polishing.

Swarovski Crystal
Austria
Marks/Description: none. Swarovski is one of the largest glass cutting factories situated in Wattens, a small tyrolean village in Austria. The remarkable precision of the facets enables each piece to reflect light perfectly.

Ca. 1980
Value: $300 (Five-piece set)

INSIDER ADVICE

Although glass is fragile it is also durable, and evidence has shown that it lasts for thousands of years. Glass is of universal interest for both style and technique in creating pieces of art, and is as costly as it is valuable. My observations about glass include:

- Of all collectibles, glass is the most difficult to identify and authenticate, and is the easiest to fake.
- False signs of wear and tear are made to imitate or project age.
- Valuable glass has a "right" weight and sound.
- Pontil marks indicate approximate date of manufacture.
- Experts can conceal chips, scratches and cracks.
- Don't use ammonia for cleaning glass.
- Ventilate showcases where glass is kept.
- Insist on comparing an item with an authentic piece before making an expensive purchase.
- Look for signatures, marks and diamond point engravings.

THE METAL MOUSE

THE VIENNA BRONZE MOUSE

The Mathias Bermann Vienna Bronze animals that are made today are hand cast and hand painted. After his death, a great granddaughter continued producing these figurines the same way. Bermann is considered by many authorities to be the father of the Vienna Bronze Animals.

This company, now owned by Ernst Hrabalek, continues to produce selected animals by skilled engravers and sculptors. Each figure is special, amusing and delightful. Attractive shapes, coloring and sense of humor make these pieces a collector's delight. The colorful scenes with mice taking on human form no doubt originated from the world of European fairy tales. Ernst Hrabalek, author of *Wiener Bronzen*, states that it is most difficult to complete an entire mouse figure orchestra.

During World War II, many bronze figures disappeared. Those that did survive the air attacks served as a nucleus, and production continues. The Bermann "Vienna Bronzes" are considered by many to be the "Rolls Royce" of miniatures.

In the 1830s, a new word was coined. Artists who turned out small animal bronzes were called "animaliers." A major part of our collection is made by Vienna Bronze artists. Bronze sculpture was a leading art form in the Renaissance period.

Vienna Bronze Mouse with Radish
1³/₁₆in (3.02cm)
Austria
Fritz Bermann Co.
Marks/Description: "FBW" in circle
Ca. 1935
Value: $135

THE COPPER MOUSE

Many collectors prefer the glowing reddish color of copper to the look of brass or gold. Antique copper pieces are quite plentiful from the middle of the 18th century onward. Good coppersmiths often specialize in engraved and embossed pieces. Copper is easily identifiable by color and malleability.

THE BRASS MOUSE

Brass is an alloy of copper and zinc. Its shiny appearance makes this metal clearly distinguishable. Some pieces are hollow, some solid, some smooth, and others hammered. When you learn that a certain mouse that you've collected is made from a special process, this item now takes on an added interest and understanding. Brass is easier to work with than any other alloy since it is softer and more easily melted, and can be more easily cast into molds because of its hardness.

GOLD AND SILVER MICE

Early gold and silver is not hard to find today. However, because of market fluctuations, we rarely find a gold or silver piece of jewelry at what we consider to be legitimate prices. Often, dealers may try to portray an exaggerated sense of an item's value when, actually, there is no shortage of fine pieces. It will take careful looking to locate a good mouse at pre-inflationary prices. Look for the "K" stamp on gold pieces.

Some of the mice we have are plated (silver on copper), some are sterling, and some are solid silver. Look for the legitimate hallmarks and identification marks on each piece. Occasionally, unscrupulous dealers may use forged or counterfeit marks to suggest that the items are genuine. Reproductions are all too common. Silver is probably the easiest metal to recognize — simply turn it upside down and look for the marks. Frequently a group of small hallmarks is impressed:

- A lion means solid silver (sterling).
- A king or queen means English made.
- A letter gives the year.
- A leopard's head means the piece is made in London.
- A king's head or a lion do not appear on American pieces.
- 925 parts out of 1000 can be considered solid or sterling silver in the United States and England.

Mouse
Cartier, NY
Marks/Description: sterling silver; Cartier
Ca. 1980
Value: $225

Fat Gold Mouse
USA
Marks/Description: 14K gold; ruby eyes
Ca. 1992
Value: $700

HALCYON DAYS ENAMELS

The rare and beautiful enamels that are made today by the Halcyon Days Enamels Company are direct descendants of 18th century England. Craftsmen hand-paint their enamel designs at Bilston, and these enamels are prized worldwide. The Halcyon Days Enamel mark can easily be identified and authenticated. This mark is a guarantee of quality.

Enamel Mouse Box
2³/16in (5.56cm)
England
Halcyon Days Enamels
Marks/Description: Halcyon Days
Enamels; wreath and crown official seal
Ca. 1985
Value: $190

THE PEWTER MOUSE

Pewter was developed and manufactured because gold and silver were always so expensive. Pewter, an alloy, grey in color, contains mostly tin and is usually cast because it is not as malleable as other metals. Therefore, pewter pieces are usually engraved and embossed, and because the metal is still quite plentiful, pewter mice are easy—and relatively inexpensive—to collect.

Pewter marks are found on the bottom of each piece. These stamping marks are called touch marks. These marks are comparable to the hallmarks in silver since they identify the maker and the town and the year. Because pewter is usually of superior design and workmanship, it is worth collecting. The quality of the metal is reflected in its appearance. As a general rule, the more it looks like silver, the finer the pewter. Unfortunately, some good pewter was never marked. Marked American pewter of 1870 is invaluable.

The pewter mice in our collection were made by the following manufacturers:

RAWCLIFF PEWTER

Traditional quality and craftsmanship go into every piece of Rawcliff Pewter. Since this pewter is 94% tin, intricate details can be reproduced with total accuracy. The Rawcliff Corporation of Providence, Rhode Island has been in the metal business for the past four generations. Outstanding skill, traditional quality and craftsmanship go into every pewter piece Rawcliff manufactures.

Mouse Box
1⁵/8in (4.13cm)
England
Halcyon Days Enamels
Marks/Description: inside hinged cover is a picture of a field mouse in a wheat field. On base: Halcyon Days Enamels: England with trademark
Ca. 1985
Value: $185

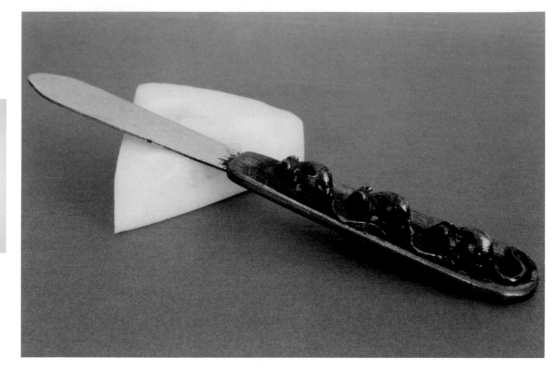

*Three Pewter Mice
on Cheese Knife*
USA
Marks/Description:
Metke: 1997
Ca. 1977
Value: $35

HERITAGE PEWTER

The use of pewter dates back to the Roman days, when it was often substituted for wood or pottery. The early pieces were simply made but durable. Today, because of the composition of the metal, fine detail can be produced. Each Heritage piece by B & J Manufacturing is handcrafted and finished to perfection with personal care. Early pewter pieces are expensive and highly prized.

With respect to the previously described mediums, I offer the following observations:
- When touch marks are faint, try using an ink eraser to bring them out more clearly.
- Don't buy old, worn or pitted pewter; it cannot be restored.
- Old brass has a silky texture due to polishing. Patina enriches brass and copper.
- The markings "Triple-plated" or "Quadruple-plated" refer to the amount of silver deposited on the base metal.
- Hallmarks are sometimes purposely worn down to make the piece look like it's very old.
- If silver is not hallmarked, its value is considerably reduced.
- Be advised that wax and lacquer will prevent normal patina from forming.
- Pure silver is too soft for practical use.
- Don't rub or polish silver too hard when cleaning, especially hallmarked sections.
- Signed work is more valuable, but this is no guarantee of authenticity.
- Gold can be plated, washed, rolled, and filled. Look for the marks.
- Brush gently when cleaning jewelry.
- Check stone settings and clasps on all jewelry.
- Separate jewelry when packing. Abrasion can cause damage.

THE PORCELAIN MOUSE

Collecting porcelain mice, like collecting any antiques, is a very personal matter. Porcelain mice are probably our favorite type to collect.

"China" has become a general term. It really should be the common name for all porcelain objects, since the word "china" was the country of origin. Porcelain is hard, white, thin, and lightweight. It is translucent when held up to light, whereas earthenware is soft and opaque. Both, of course, are made of clay. The clay material used is what makes up the difference between the two forms. Different colors and temperature used in the manufacturing process alters the composition. This accounts for the dissimilarity between porcelain pieces.

As a general rule, most European manufacturers mark their porcelain with some type of identifying mark. Usually a symbol, such as a shield or a crown or a signature, can be found. Most marks are straightforward. The crossed swords of Meissen and the interlaced "L's" of Sevres are most copied. Since many marks are applied by hand, it is very rare that you'll find two exactly alike. Therefore, the mark and other criteria have to be examined carefully before making a true certification.

Be aware that, as old mice become more scarce, you will find more imitations being manufactured. Unfortunately, it is extremely difficult to differentiate between the old originals and some newly-made reproductions.

Today, porcelain statues are the most sought-after items at antique shows. A friend of mine who is a dealer reports that he hardly has a show go by in which he fails to sell most of his porcelain animal figures. He also tells me that many of the pieces that are purchased are not antiques in the traditional sense of the word, since many are less than one hundred years old. On earthenware the identifying marks are either made with a small metal point, impressed in the clay, or by painted or printed marks. Both methods may be regarded as "genuine marks" on old pieces provided they are under the glaze. The marks of the larger and well-known companies were often copied by their rivals. No marks were used on early American pottery.

The porcelain mice in our collection were made by the following manufacturers:

CYBIS COMPANY

The Cybis Company of Trenton, New Jersey is America's oldest existing porcelain studio. Artists at Cybis follow the traditions established by Boleslaw Cybis, the founder, and create fine porcelain sculptures that are prized by art collectors around the world. Done in the fashion of the old world studios and known for their beauty and increasing valuation, each of their sculptures are individualized works of art as each artist mixes his or her own paints for decorating a sculpture.

Porcelain Cybis Mouse with Acorns
6in (15.24cm)
Trenton, NJ
Cybis Porcelains
Marks/Description: Cybis signature; letter R; "Maxine" the Dormouse; Animal Kingdom and Woodland Collection
Ca. 1977
Value: $315

BING AND GRONDAHL

The famous Bing and Grondahl factory has been making fine porcelain in Copenhagen, Denmark, since 1853. During this period, overglaze was the primary procedure. In 1866, the unglazed blue decoration began. Figurines are still being made the same way today. Recently, Royal Copenhagen and Bing and Grondahl merged and are considered the leading creators of the world's best Danish porcelain.

BENNINGTON

In 1703, John and William Norton started making earthenware in Bennington, Vermont. This was known as the Norton Stoneware Company. In 1849, the name was changed to the United States Pottery Company. Bennington figures were carefully molded with a rich velvety sheen. Fine Bennington is rare and was rarely marked. The factory closed in 1858.

Card Tray with Two Mice
5in (12.7cm)
Denmark
Bing and Grondahl
Marks/Description: Dahl Jensen: Made in Denmark: B & G #1562W & Castle (The mice have pink eyes and tails that wrap around entire tray.)

Ca. 1950
Value: $290

Porcelain Mouse on Trap
2½in (6.35cm)
Scotland
Schmid Border Fine Arts
Marks/Description: Ayers, SMC, BFA
Ca. 1981
Value: $85

SCHMID BORDER FINE ART

The process involved in creating Schmid Border Fine Art sculpture porcelain is extremely time-consuming. The countless stages in the production process require time and patience. Often, a thousand hours or more of work are required to create a primary figure. This family-owned company was founded in 1931 by John Hammond. The Walt Disney characters, Beatrix Potter, and the Hummel figurines are just a few of the exclusive sculptures distributed by Schmid.

HEREND

This porcelain company was established in Herend, Hungary, by Moritz Fisher in 1839. The pieces are referred to as Herend porcelain, and the company is still in operation.

THOMAS MINTON

Thomas Minton founded his company in 1796. Porcelain was not produced in any great quantity in its early years, however, after 1825, quality and quantity increased. Minton made some of the best porcelain in England during the Victorian period. The company is still manufacturing fine quality porcelain today. Marks include the letter "M" and the name "Minton" impressed.

Porcelain Mice
Hungary
Herend Corporation
Marks/Description: blue and white coloring; hand painted; #205 imprinted; #5340/VHB impressed
Ca. 1995
Value: $175

Minton Majolica Vase
7in (17.78cm)
England
Marks/Description: Minton (under glaze). Blue-green-yellow vase with bird and mouse. Generally speaking, the term "Majolica" means any glazed pottery. Pottery is glazed with tin enamel thus concealing the true color of the clay. Pottery like this is manufactured in many countries. Old Majolica is best because of its beautiful design. Majolica pieces are regarded as genuine if the marks are under the glaze.
Ca. 1880
Value: $3,700

DRESDEN CHINA

Dresden is any china made in the town of Dresden, Germany. The Dresden factory was established in 1709 where stoneware resembling porcelain was produced. Later, the company moved to Meissen, where it is still manufactured. Many inexperienced collectors have purchased old Dresden imitations. Meissen pieces usually have crossed-swords marks and are listed under Meissen.

SEVRES

The King of France granted Sevres authority to manufacture fine quality porcelains since the mid-1700's. Sevres porcelain, both the soft-paste and the hard-paste, are absolutely translucent and flawless. One of the most forged marks in antiques is the twisted lines with a center letter imitating the Sevres mark. Rich colors often enhanced the hand-painted art of Sevres porcelain. In the early years heavy gold decoration enriched the Sevres porcelain making these pieces hard to find and very expensive today.

BOEHM

In 1950, the Edward Marshall Boehm Co. was established in Trenton, New Jersey. Although Mr. Boehm died in 1969, the firm has continued to design and produce its remarkable porcelain under the leadership of his wife Helen. Ed Boehm formulated his hard-paste porcelain himself, which required special handcrafting skills. Limited and non-limited edition pieces are produced. Once a limited edition is complete, the original mold is broken and the edition closed. The production of a single piece of sculpture requires more than twelve steps before its final inspection. Today, one can find Boehm porcelains in the White House and in museums around the world. The work of the artists and craftsmen of the Boehm Studios—in every sense of endeavor—can never be matched by machinery. What they have accomplished already, and what they have challenged themselves to do in the future, is a rare and remarkable talent.

Porcelain White Mouse Preening
Trenton, NJ
Marks/Description: Edward Marshall Boehm, USA, Pink eyes and ears, Hallmark R.P. #513

Ca. 1960
Value: $850

CONNOISSEUR THISTLEDOWN STUDIO

The Connoisseur Thistledown Studio was founded in England in 1978, and relocated to New Jersey one year later. Today, their small studio creates exclusive and different works of porcelain art. Thistledown sculpture is handcrafted in a limited edition of one hundred pieces. The artists of Connoisseur are known for creating unique, exceptional studies of wild animals, flowers, birds and many historical subjects.

NYMPHENBURG

The Nymphenburq Porcelain Company was established in 1753 and is still producing today. A crown with a checkered shield below, and a crowned "CT" with the year are the modern marks.

Above:
Porcelain Marriage
England
Connoisseur of Malvern
Marks/Description: "The Wed" edition 11/100; fine bone china; Thistledown Collection; Brielle Galleries Exclusive
Ca. 1990
Value: $750

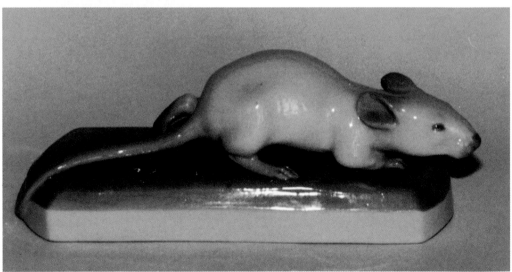

White Porcelain Mouse with Pink Eyes, Ears, and Tail
4½in (11.43cm)
Nymphenburg, West Germany
Marks/Description: stamped #475 and #27 and seal; Nymphenburg
Ca. 1950
Value: $375

ROYAL CROWN DERBY COMPANY, LTD.

The Royal Crown Derby Company, Ltd. started to manufacture porcelain in England in 1876. Since 1921, the company's mark is the crown, its name, and the words "Made In England."

ZSOLNAY

The Zsolnay brothers created original and creative porcelain pieces. Although no trademark was used in the early period, their work was recognizable by its ornamental and glazed appearance. In 1878, a blue mark signifying the towers of the cathedral at Pecs was used. New Zsolnay figures with their green-gold finish have shown fairly consistent appreciation and are available at retail establishments.

Royal Crown Derby Mouse
2¼in (5.72cm)
England
Marks/Description: Royal Crown Derby English
Bone China
Ca. 1980
Value: $115

Zsolnay Mouse
2½in (6.35cm)
Pecs, Hungary
Marks/Description: tower
trademark and #9504
Ca. 1880
Value: $375

MEISSEN

The Meissen factory was the first in Europe to produce hard-paste porcelain. The factory was near the town of Dresden. From the early 1700's through the present Dresden has produced fine porcelain and stoneware. No identification marks were apparent on the early Meissen pieces. Occasionally, crossed-swords with dates appeared, and a dot was added between the swords in the early 1920s. Counterfeiters have imitated Dresden mark of the crossed swords and dot. Sharp and bright colors and durable porcelain is a sign of original Meissen.

Porcelain Meissen Mouse
2³/₁₆in (5.56cm)
Germany
Marks/Description: #1120 & F8502; incised with trademark

Ca. 1960
Value: $525

ROYAL COPENHAGEN

The oldest studio in Denmark is called Royal Copenhagen. L. Fourier directed and manufactured porcelain that was soft-paste having excellent quality features. Some of the best hard-porcelain produced today is made by Royal Copenhagen. Figures, especially animals, having quiet colors are most noteworthy. The Copenhagen mark usually has three wavy lines.

Porcelain White Mouse on Walnut
2½in (6.35cm)
Denmark
Royal Copenhagen Porcelain Factory, LTD.
Marks/Description: crown; Denmark; #511; CA. 1922; Stoneware, Faience
Ca. 1960
Value: $200

DOULTON COMPANY

George Tinworth, one of the most famous sculptors of that time, joined the company in 1870, revitalizing it. Either his monogram or his signature with the factory name and date is indicative of genuine Doulton porcelain. The Burslem branch was best known for its ornamental and appealing bone china. Since 1902, the "Royal" mark has been used. Royal Doulton turns out a singular piece each year that can only be acquired from Doulton during that year. One of its most famous pieces is the Toby Mug. Today the company still produces lambeth faience, and flambe figures. The Beswick company was absorbed by Doulton in 1969, but continues production under the Beswick name.

Mouse Pushing Wheelbarrow with Blue Vase
2½in (6.35cm)
London, England
Doulton Stoneware
Artist: George Tinworth
Marks/Description: official impression on base. George Tinworth was born in London on November 5, 1843, and died there on September 10, 1913. He worked as a modeler and decorative sculptor. He was noted for his excellent terracotta amusing animal groups in stoneware. All of his figures bear his signature or monogram, factory mark, and usually, a date.
Ca. 1885
Value: $1,200

Two Porcelain White Mice
3¾in (9.53cm)
Selb, Bavaria
Rosenthal
Marks/Description: K. Himmelstoss; crown with crossed swords; stamped K188, Rosenthal, Selb, Bavaria

Ca. 1920
Value: $225

ROSENTHAL

The Rosenthal Porcelain factory in Selb, Bavaria was established in 1879. Philip Rosenthal painted his own designs on pieces that he purchased from other companies when he first established himself. The company is still in business and you can find the Rosenthal signature on the back of each piece.

NATURE"S HERITAGE

Nature's Heritage produces cold-cast porcelain. Combining porcelain with resin allows the artists and craftsmen to fashion exceptionally fine detail and superior design to reproduce nature's appearance. Each piece is produced entirely by hand, and only the finest materials and paint are used. Therefore, no two pieces are exactly alike.

Porcelain Brown and White Mouse in
Green Apple
2¾in (6.99cm)
Staffordshire, England
Holland Studio Craft
Marks/Description: hand painted by "A.M.";
Nature's Heritage by Holland Studio Craft
Ca. 1987
Value: $85

LLADRO

The classical procedures that produce the distinct elongated design was produced by the Lladro Brothers in Spain in 1951. Their porcelain and stoneware are made in limited and non-limited editions.

Porcelain Cat and Mouse
3in (7.62cm)
Spain
Lladro Bros.
Marks/Description: Ref 05236; Gatito Pasmado
Ca. 1985
Value: $95

LENOX

Lenox has made decorated pieces of porcelain in its factory in Trenton, New Jersey since 1896. The company today produces thin porcelain pieces often with a pearly glaze and soft creamy colors. Lenox is known for its transparent appearance and warm colors.

LIMOGES

The Limoges porcelain factory was established at Limoges, France in 1771. The king purchased it in 1784. After this, it served as a branch of Sevres. Modern porcelains are still being made at Limoges.

Three Blind Mice
France
Limoges
Marks/Description: Aztozia, Peint Main,
Limoges, #237, "LP"
Ca. 1990
Value: $125

HAVILAND

David Haviland started his porcelain factory in Limoges in 1840 specifically to ship his porcelain to the United States.

NORITAKE

Noritake porcelain was made in Japan after 1904. Usually the white porcelain pieces were exported to other countries to be decorated. Although the firm is still working today, it sustained a great deal of damage during World War II. Pieces with the letter "M" in a wreath for the Morimura Brothers in New York are the best Noritake pieces.

HUTCHENREUTHER PORCELAIN COMPANY

In 1814, the Hutchenreuther Porcelain Company of Selb, Germany, was established. This company still makes fine quality hard-paste porcelain today. The lion and insignia appear most of the time even though the mark has changed.

COALPORT

The Coalport Works in England was established in 1795. John Rose started this company. He imitated Dresden and Sevres and frequently counterfeited their marks in order to deceive buyers. This caused considerable confusion. The chief mode of decoration is still the under painting process. The Rose family retired from manufacturing at which time, in 1862, the Wedgwood Company took over. With some variations, the crown mark has been used since the late 1880s and is still being used. "Coalport" in script is marked on their sculpture.

Noritake Bone China White Mouse
4½in (11.43cm)
Japan
Marks/Description: Noritake; bone china; Nippon Toki Kaisha

Ca. 1985
Value: $65

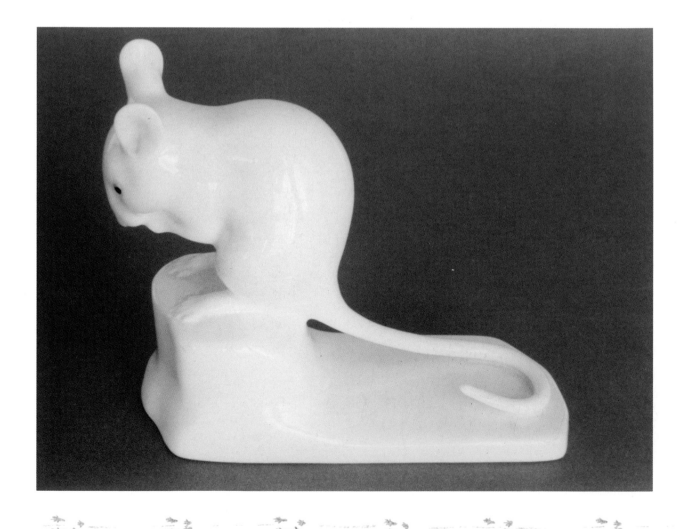

White Mouse
4½in (11.43cm)
Germany
Hutschenreuther
Marks/Description: lion facing left;
Germany; #181
Ca. 1935
Value: $125

White Porcelain Mouse with Brown Eyes
2³/₁₆in (5.56cm)
England
Coalport
Marks/Description: crown; Coalport
Ca. 1970
Value: $22

WEDGWOOD

One of the world's most successful establishments was started by Josiah Wedgwood in 1759. The factory is still operating today and produces jasper and basalt. It is often difficult to read the impressed marks that are used for identification. Therefore, look carefully at the bottom of each piece to determine that you are getting a legitimate Wedgwood porcelain. Three letters indicate the month, craftsman, and the year of production. The official seal was put on the piece from 1860 to 1930, but after 1930 the actual date and the official seal were put on.

Wedgwood Blue Vase with Eight White Mice
8⅝in (21.9cm)
England
Wedgwood
Marks/Description: Wedgwood, England;
official seal & Z 5179 on base. Porcelain vase and artwork designed by James Hodgkiss.
Ca.1902
Value: $1,200

GOEBEL

The Goebel factory was established in Germany in 1871. The firm is still manufacturing. Because the figurines are not made in a limited edition, they can be reintroduced by the company at any time.

Three Porcelain Mice
2-2³/₁₆in (5.08-5.56cm)
W. Germany
Goebel
Marks/Description: #35795, Goebel, W. Germany
Ca. 1985
Value: $54 (set)

HARMONY KINGDOM

Crushed marble and resin are the ingredients that create the look and feel of marble when Harmony Kingdom's craftsmen create their figurines. These remarkable miniature subjects are both elegant and whimsical and often imitate the style of the original Japanese netsuke. Many Harmony Kingdom pieces are available in limited editions from the Harmony Ball Company.

Porcelain Mouse Box
1in (2.54cm)
China
Harmony Kingdom
Marks/Description: "Field Day"; handcrafted in China; trademark (HK)
Ca. 2000
Value: $40

With respect to porcelain figurines, I offer the following observations and suggestions:

- Marks can be removed as well as added or changed.
- If there are several parts to an item (e.g., pot/cover) make sure all parts are original.
- Use ultraviolet light to determine if a piece was repaired.
- Hard-paste porcelain can't be scratched with a piece of metal. Soft-paste porcelain can easily be scratched so don't experiment. (If you select the wrong piece it's going to cost you!)
- Unfortunately, many so-called "hand painted" porcelain pieces are frequently produced on an assembly line.

Marks Miscellany

- Nymphenburg porcelain is almost always marked with an impressed shield. The mark, which is very conspicuous, is always cross-hatched diagonally.
- Entwined "L"s with a center letter represents Sevres.
- The Copenhagen mark usually has three wavy lines in underglaze blue.
- Meissen uses crossed swords.
- Early Coalport pieces are not marked. Later some were marked with the script "Coalport".
- Minton marks were based on the entwined "L" of Sevres. During the mid-nineteenth century,
- the mark was changed to include the letter "M" and the name "Minton" impressed or transfer printed.
- Cybis usually has an embossed and script "Cybis".
- Bing and Grondahl use the initials "B & G" and a castle.
- Bennington rarely used a mark.
- Schmid Border Fine Art uses a paste-on official seal and the name of the artist impressed on the porcelain.
- Dresden pieces are usually marked with the words "Dresden" and "Germany" with the letter "S" below a crown.
- Herend porcelain has the name "Herend" impressed and the word "handpainted" with a crown and paintbrushes.
- Limoges uses the name "Limoges" as part of its mark.
- Wedgwood's mark is a vase, the name "Wedgwood," a star, and "England."
- "Goebel" under a crown is the mark for Goebel.
- The name "Royal Doulton" appears after 1902.
- Rosenthal uses a crown over its name as a mark.
- Boehm uses a horse head and its name.
- The mark of Connoisseur is its name and trademark.
- Lladro uses its name under its trademark.
- Haviland uses several marks. "Haviland & Co." is one.
- Noritake uses its name with "M" in a wreath as a mark.
- On most pieces, the mark for Hutchenreuther is a lion and its name.
- Lenox uses several marks. Check earlier marks for best source.
- The mark for Holland Studio Craft is "HSC", and the year impressed.
- The Nymphenburg marks include a checkered shield topped with a crown, and a "CT" with the year.

Building A Better Mousetrap

or, how to display your collection

Part of our mouse collection is displayed in
this side by side oak desk/bookcase.
1895

If you are proud of your collection—and you should be—you are going to want to show it off. That's why you should be just as careful in choosing your display furniture as you were in selecting your treasures. Every mouse seeks a safe, comfortable home, so when you select furniture, whether you purchase a ready-made cabinet, or a made-to-order expensive curio unit to house your collection, you want it to be in good taste, without detracting from the pieces within.

Although you want something attractive, don't allow the furniture to be the focal point of your collection. It should be appropriate looking, but don't let it steal the show. You also want the mice to be arranged in such away that one piece doesn't compete against another or depreciate others. If you have a knock-out figurine, let it stand on its own. Don't surround it—or even put it next to—inferior pieces. We enjoy showing our collection to other mouse collectors, friends, and relatives, and so we've arranged our mice to make them more attractive looking and also to enhance the room.

In order to accommodate so many items, we tried several designs to make the collection more striking. First of all, the pieces were arranged in groups; it gives the collection a certain rhythm and more eye appeal. Where possible, closed cabinets were electrified. (I find that a 40 or 60 watt bulb is quite sufficient to display the figurines in our cabinets.) Some that I designed myself have a mirror backing so that the figures can be seen from different angles. Glass mice were grouped together. We've secluded the Vienna Bronze pieces in a separate cabinet, and the porcelain critters in still another compartment. Orchestras are so unique that we decided to place them all to themselves. The large pieces were placed in a special oak desk (more about this desk below) which I electrified myself since the shelving area was better for large items (e.g., pitchers, vases).

Don't line up your mice like wooden soldiers. Instead, let your placement create the impression that these pieces are interacting with one another. Make the exhibition arresting. When you arrange different statues by size, side by side, you want them to be appealing to the eye. We organized the mice in attractive appointments but made sure the furniture and fixtures did not outshine the assemblage. Realize that, almost without trying, in just a few years you can accumulate several thousand dollars worth of antiques and collectibles. These subjects deserve top honors. Like I said, if you purchase good items, they will invariably increase in value in time. It's not as important to have a large number of subjects just to fill up a showcase, as it is to have quality. A few superior pieces are far better looking and have more value than a cabinet overflowing and cluttered with many unimportant pieces.

If you can afford to have a display section custom made, by all means do so. The results are really evident. This way the unit is made specifically for you, and it shows. Friends of ours who collect ivory, (mostly Japanese netsuke) for example, display their figurines in wood cabinets that are painted black. The ivory shows up so beautifully against this ebony and black velvet background.

Display cases often serve a dual function. When some friends decided that they wanted to separate their living room from their dining area, they called in a carpenter to create a wall that would incorporate twelve of their Edward Marshall Boehm porcelain birds. Glass cubicles were placed inside the wall with hidden lights in each compartment. The result is dramatic— breathtaking—especially at night.

Lighting, of course, is especially important. In a custom wall unit, you can highlight shelf by shelf or section by section, and you can utilize directed ceiling lights. As I mentioned before, most of our collection is housed in glass enclosed cabinets. However, we could not resist the antique (1885) oak desk, with its attached cabinets on either side, which proved to be the perfect carrier to display our larger collectibles.

Other mice are arranged in separate categories. Colorful porcelain pieces are set apart from the others, and singers and dancers and musical figures are located near the orchestras.

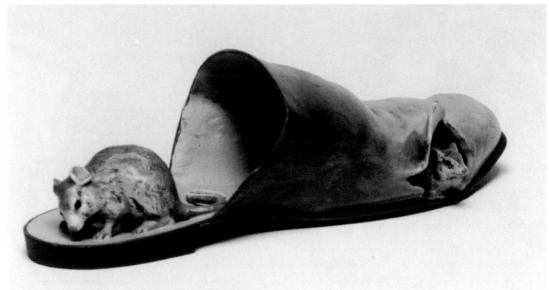

Above:
Orchestra
Austria
Marks/Description:
Geschotzt Austria
and letter "B" in
trademark
Ca. 1930
Value: $3,500

Vienna Bronze Mice on Slipper
6in (15.24cm)
Austria
Fritz Bermann Co.
Marks/Description: "FBW" in circle
Ca. 1930
Value: $490

The silver exhibition is grouped together away from the crystal pieces, which are arranged in their own section. Netsukes and ivory carvings have one shelf, and pewter models take another. Here and there we have several "mix 'n match" pieces that compliment each other. It gives the collection a certain cohesiveness.

Any ideas or props at your disposal should be used to enhance your collection. Don't, however, allow your treasures to be spoiled by a well-meaning gift. Usually a copy or a replica stands out like black ink on a white shirt.

Most important! If you purchase rare and valuable antique pieces be sure to catalog them, take pictures of them, and insure them. Set up a filing system. List the price, purchase date, and frequent appraisal values.

Have fun with your collection. Be creative. Don't forget, the assemblage didn't grow overnight. You have taken your time in assembling a collection, and the manner in which it is presented should be reflective of, and complimentary to, the care and thought that you dedicated in making each acquisition.

Do children and collections mix? We think they do. Of course you have to make your valuables child-safe. Still, we think it's important for your children—or grandchildren—to have the opportunity to enjoy your collection, too. When our grandchildren come to visit, they know that the mice in the cabinets are off-limits. But we always display some of our less-valuable, non-breakable mice on our coffee tables and end tables, and our grandchildren are free to handle and admire them. Personally, I think this is a good hands-on opportunity for young children to learn how to appreciate something of value.

Necklace
Japan
Marks/Description: Netsuke mice and jet black beads
Ca. 1975
Value: $250

Ivory Netsuke
¾in (1.91cm)
Japan
Marks/Description: styled necklace with 14K mounting
Ca. 1950
Value: $75

Netsuke: Hand-Carved Ivory Mice on Skull
1½in (3.81cm)
Japan
Marks/Description: Japanese signature; Netsukes were made of wood, ivory, porcelain and metal. Averaging between one to three inches tall, Netsukes served as small fasteners like a button, which helped prevent a hanging object from falling off. These standing free art forms are treasures. Very few old pieces are signed.
Ca. 1900
Value: $175

Four Mice on a Mouse Netsuke
1½in (3.81cm)
Japan
Marks/Description: Japanese signature
Value: $155

Ivory Netsuke
¾in (1.91cm)
Japan
Marks/Description: face of Netsuke mounted as ring with 14K gold wire
Ca. 1960
Value: $75

Mickey, Idol of Millions

Plane Crazy Porcelain Disc
3¹/₈in (7.94cm)
Japan
Marks/Description: Disney Land, Walt Disney
World, Walt Disney Productions
Ca. 1928
Value: $40

According to the stories that have been repeated so many times, the birth of Mickey Mouse has now become somewhat of a legend. From what I've read, Walt Disney fashioned the idea of a mouse for a cartoon while he was riding on a train ride from New York to Hollywood. According to history, it seems that Walt Disney captured a mouse in his Kansas City office. It is written that he kept it in his waste-paper basket on his desk, and that after a period of time he was able to tame it, and later feed it by hand. Originally, his wife was going to call the mouse Mortimer, so they say. However, it was finally decided to call him Mickey, and the name stuck.

This was the beginning. Mickey's first cartoon was called *Plane Crazy*, which was based upon Lindbergh's flight. This was followed by *Steamboat Willie* a great success. Did you know, Mickey was the first cartoon character to talk? Mickey won the public over and became one of the biggest stars in Hollywood. In my view, Mickey will never grow older as long as the writers and artists continue to develop their delightfully amusing stories. Never knowing what Mickey is up to is what captures the audience. Incidentally, Mickey was born in 1928 and as you can see has never gotten a day older.

Today the creators of Mickey Mouse have made him an international figure, and possibly one of the biggest commercial endeavors of all times. All around the country you can find Mickey Mouse clubs. Some groups publish their own magazines, and even have private auctions. And to think this all started with a cheese-eating furry critter.

Mickey and Minnie Plaques
Japan
Marks/Description: Disney Trademark
Ca. 1990
Value: $15 (pair)

Disney book: *Disney Babies At The BIG Circus*
USA
Walt Disney Product
Marks/Description: A book about opposites;
match the pictures that show things that are
opposite; written by Rita D. Gould
Ca. 1987
Value: $5

Mickey and Minnie Plate
7½in (19.05cm)
China
Marketed by Gibson Overseas, Inc.
Alhambra, CA
Marks/Description: dishwasher and microwave
safe; soup plates and dinner plates also available
Ca. 2000
Value: $6, $10, $12

Above:
Jigsaw Puzzle
USA
Walt Disney Product
Ca. 2000
Value: $10

Left:
Lunch Box
China
ASC Co., Lancaster, PA
Marks/Description: 17603; #41
Ca. 1985
Value: $15

Mickey Cookie Jar
China
Marks/Description: none
Ca. 1990
Value: $20

Jumbo Mickey
26in (66.04cm)
China
Fisher-Price Inc.
Marks/Description: Disney
Trademark
Ca. 1992
Value: $35

*Socks, One Pair White/
One Pair Black*
Taiwan
High Point Knitting, Inc.
N.Y., N.Y.
Marks/Description: 80%
acrylic and 20% Nylon;
Disney Trademark
Ca. 2001
Value: $3 (per pair)

Mickey Sweatshirt
China
Marks/Description: #1928; classic legend
Ca. 2001
Value: $10

Official Mickey Mouse Hat
USA
Walt Disney Company
Marks/Description: Made in U.S. by Benay Albee
Ca. 1980
Value: $10

Above:
Mickey Mouse Sweatshirt
Japan
Marks/Description: front and
back of same sweatshirt;
Walt Disney trademark
Ca. 1995
Value: $12

Right:
Umbrella
Japan
Marks/Description:
Walt Disney trademark
Ca. 2001
Value: $7

Mickey Mouse Eyeglass Cases
China
Distributed by Marchon
Marks/Description:
Disney trademark
Ca. 1995
Value: $5 each

© Disney

Counterfeit
This is an imitation. This photo of Mickey has been selected to make you aware. Don't ever purchase an expensive figure, or for that matter, any piece that a dealer claims to be genuine unless you check the piece's authenticity. This Mickey has no Disney trademark front or back to show that it's an original.

Mickey Mouse Toy—Mickey Says
Malaysia
Milton Bradley Co.
Marks/Description: Letters "MB";
The Walt Disney Company
Ca. 1998
Value: $15

Mickey Bank
China
Marks/Description:
trademark Disney
Ca. 1990
Value: $3

Tin Box
Hong Kong
The Tin Box Co. of America, Inc., Long Island
City, N.Y.
Marks/Description: The Walt Disney Co.,
DNY-2M, SM Tall Round
Ca. 1988
Value: $3.50

Official Mickey Mouse Club Certificate
USA
Marks/Description: none
Ca. 1975
Value: $1

Mickey Thimble
Japan
Schmidt
Marks/Description: Walt Disney
character; #279-662; Snow Biz
1985 Annual Thimble
Ca. 1985
Value: $10

Left:
Baby Mickey
China
Canassa Trading Corp., Burbank, CA
Marks/Description: Disney Land, Walt
Disney World
Ca. 1995
Value: $12

Below:
Mechanical Mickey
China
Mattel Inc., El Segundo, CA
Marks/Description: battery operated;
crawls and makes noises; Disney
trademark
Ca. 1999
Value: $25

Mickey Mouse in Blue
6¼in (16.51cm)
China
Dolly Inc., Tipp City, OH
Marks/Description: Disney
trademark; polyester fibers
Ca. 1995
Value: $10

Mickey Mouse in Green
6¼in (16.51cm)
China
Dolly Inc., Tipp City, OH
Marks/Description: Disney
trademark; polyester fibers
Ca. 1995
Value: $10

Baby Minnie Mouse
6in (15.24cm)
China
Dolly, Inc. Tipp City, Ohio
Marks/Description: new materials only; polyester fibers

Ca. 2000
Value: $8

Minnie Mouse
9½in (24.13cm)
China
Marks/Description: Walt Disney World; polyester fibers; head, arms and legs all movable
Ca. 1995
Value: $18

Minnie at the Piano
10⅝in (26.99cm)
Japan
Carousel Leaf Co., Lake Forest, IL
Marks/Description: plays musical notes when pressed; Disney trademark

Ca. 1995
Value: $25

Baby Minnie Mouse
9in (22.86cm)
China
The Walt Disney Co., Applause Co.
Woodland Hills, CA
Marks/Description: synthetic fibers;
do not immerse; item #33762
Ca. 1995
Value: $9

Minnie Mouse
China
Canasa Trading Corp., Burbank, CA
Marks/Description: Disneyland,
Walt Disney World
Ca. 1995
Value: $15

Handbag
China
Pyramid Handbag Co. N.Y., N.Y.
Marks/Description: "Mickey's Stuff"
Ca. 2001
Value: $12

Leather Patch
Korea
The Walt Disney Co.,
The Zephyr Group, Inc.
Marks/Description: Torrance, CA; genuine leather

Ca. 1995
Value: $10

Minnie Wristwatch
1⅝in (4.13cm)
China
Marks/Description: Lorus Quartz, Japan parts
Ca., 1999
Value: $15

And Finally . . .

No matter what you collect, whether it's for your own enjoyment or for monetary purposes, investments in legitimate antiques frequently prove to be sound investments. However don't buy an item because it's a bargain or because it's old. Antiquity, by itself, doesn't guarantee that the piece is valuable. And you had better like your treasures—and plan on keeping them for a long time—because if you should ever want to sell them (unless you are going to make a career out of selling or dealing) the chances are few and far between that you'll find a buyer to purchase the items at the price you paid for them.

Some dealers consider items over sixty years old to be antiques. Others state that to be an antique an object must be at least one hundred years old. You will have to use your own judgement and make your own decision. Regardless of what you decide, remember if you purchase good quality, interesting, and contemporary pieces today, their high quality will ensure that someday they will be called antiques. Hang in. The collecting of rare and curious mice of the past are possessions of permanent pride.

now . . . the rest of our collection!

Brass Mirror
5¹/₈in (13.02cm)
Japan
Marks/Description: mouse holding
what appears to be a globe; other
side very highly polished
Ca. 1880
Value: $125

Solid Brass Inkwell
5³/₈in (13.65cm)
Austria
Marks/Description: none; cat
chasing mouse in cage
Ca. 1900
Value: $425

Three-Piece Band
USA
Marks/Description:
stamped Reversal
Stationary Corp.,
1974, #805
Ca. 1974
Value: $90

Brass Shoe with Two Mice
Austria
Marks/Description: none
Ca. 1990
Value: $150

Brass Victorian Stamp Box
2in (5.08cm)
Austria
Marks/Description: none
Ca. 1890
Value: $275

Personal Door Knocker
USA
Made this myself. Purchased a plain antique door
knocker and mounted this brass antique mouse.
Ca. 2002
Value: Not For Sale

Solid Brass Figure
2³/₈in (6.03cm)
Japan
Marks/Description: several
Japanese symbols
Ca. 1995
Value: $45

*Bronze Cat and Mouse
Inkwell*
5in (12.7cm)
Possibly Austria
Marks/Description:
letters "A" and "B" in a
rectangle;#281; barrel
top is hinged
Ca. 1890
Value: $370

Vienna Bronze Mouse on Plant
3³/8in (8.57cm)
Vienna, Austria
Marks/Description: Fritz Bermann Co.
Ca. 1935
Value: $300

Vienna Bronze Mice
1½in (3.81cm)
Austria
Marks/Description: Austria, Geshotzt
Ca. 1900
Value: $135

Bronze Silver-Plated Mouse on
Silver-Plated Book
3in (7.62cm)
France
Marks/Description: Odeurs De Paris; letter
"W" on binding
Ca. 1920
Value: $650

Bronze Mouse with Two Sugar Cubes on Marble Base
Marks/Description: Signed Masson. Colvis Edmond Masson (1838-1913) studied under Santiago, Barye and Rouillard and exhibited at the salon from 1867-1881, winning an honorable mention. He was an animalier.

Ca. 1885
Value: $950

Vienna Bronze Mouse with Sugar Cube
Austria
Marks/Description: none
Ca. 1900
Value: $190

Bronze Blackboard Matchbox
8½in (21.59cm)
USA
Marks/Description: letters "AB";
"A Narrow Escape"; design owned
by Juvenile MFG. Co., Dalton, OH
Ca. 1900
Value: $175

Bronze Mouse
2½in (6.35cm)
Marks/Description: initials C.B.W.;
artist signature and country of
origin illegible. This sculpture is
similar in appearance to the
porcelain white mouse preening
by Edward Marshall Boehm.
Ca. 1930
Value: $130

Bronze Mice Paperweight
2³/₈in (6.03cm)
Austria
Marks/Description: none
Ca. 1895
Value: $725

Vienna Bronze Mouse on Skis
2in (5.08cm)
Vienna, Austria
Fritz Bermann Company
Marks/Description: "FBW" in circle
Ca. 1935
Value: $225

Brass Shoe
Austria
Marks/Description: Bronze
shoe with one mouse
Ca. 1990
Value: $150

Two Bronze Mice on Bronze Lamp Stand on Marble Base
France
Marks/Description: 19 (signature G. Palrel) 20
Ca. 1920
Value: $1,200

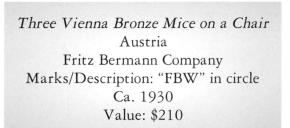

Three Vienna Bronze Mice on a Chair
Austria
Fritz Bermann Company
Marks/Description: "FBW" in circle
Ca. 1930
Value: $210

Bronze Mouse Standing
2¾in (6.99cm)
USA
Marks/Description: none
Ca. 1925
Value: $140

Vienna Bronze Pin Holders
Large Mouse 2¾in (6.99cm)
Small Mouse 1⅝in (4.13cm)
(not including length of tail)
Marks/Description: none.
Characteristic markings of Fritz
Bermann
Ca. 1930
Value: Large $170; Small $140

Vienna Bronze Dog Looking at Mouse in Rain
Barrel
Austria
Marks/Description: none
Ca. 1900
Value: $525

Vienna Bronze Mouse on Inkwell
5¾in (14.61cm)
Austria
Marks/Description: Stephanie Biscuits
stamped on lid
Ca. 1900
Value: $240

Bronze Mouse
2¼in (5.72cm)
Austria
Marks/Description: illegible
engraving under tail
Ca. 1910
Value: $145

Vienna Bronze Mouse with Fishing Pole
3³/₈in (8.57cm)
Austria
Fritz Bermann Co.
Marks/Description: "FBW" in circle. The
hand-cast bronze figures dating back 100 years
are made by the Fritz Bermann Company. They
have over 3000 patterns and all details are
painted and chiseled by hand.
Ca. 1935
Value: $175

Bronze Mouse on Fireplace Brush
5¹/₈in (13.02cm)
Marks/Description: none
Ca. 1880
Value: $750

*Bronze Cat and
Mice Ashtray*
5¹/₈in (13.02cm)
France
Colin & Co.
Marks/Description:
T. Hungre, Paris.
Louis Hungre was
born in Ecouen,
France. In 1909,
he was elected an
associate of the
Artists Francais.
He died in 1911.
Ca. 1905
Value: $1200

Above:
*Two Mice on Bronze Inkstand
with Spilled Ink*
12¾in (32.39cm)
Marks/Description: Ch. Korschmann, Paris.
Mice have ruby eyes. Charles Korschmann
was born in Brno, Bohemia on July 23,
1872. According to the records, he studied
at the Schools of Fine Art in Vienna, Berlin,
and Paris. In 1900, he was awarded a bronze
medal at the Exposition. He specialized in
marble and bronze. Listed in the *Dictionary
of Sculpture in Bronze* by James Mackay.
Ca. 1900
Value: $2,400

Right:
Bronze Ashtray
USA
Marks/Description: none; no explanation for
the mouse figure
Ca. 1920
Value: $145

Bronze Mouse
1¾in (4.45cm)
Austria
Marks/Description: Krieger. Wilhelm Krieger
worked in Hersching, near Munich in 1877. He
specialized in reproduction of animals. Listed in
Bebezit book, page 315.
Ca. 1890
Value: $225

*Two Bronze Mice
Drinking*
1⅞in (4.76cm)
Geschotzt, Austria
Marks/Description:
Geschotzt; playing
cards initials "HB"
(H. Bohatsschek)
Ca. 1939
Value: $345

Vienna Bronze Mouse with Top Hat
1in (2.54cm)
Austria
Fritz Bermann Co.
Marks/Description: "FBW" in circle
Ca. 1935
Value: $85

Vienna Bronze Mouse in Red Car
2in (5.08cm)
Austria
Fritz Bermann Co.
Marks/Description: "FBW" in circle
Ca. 1935
Value: $225

*Vienna Bronze Mouse Hunter
with Cat*
3¹/₈in (7.94cm)
Vienna, Austria
Fritz Bermann Co.
Marks/Description: "FBW" in circle
Ca. 1950
Value: $320

*Vienna Bronze
Dachshund on
Bench with Mouse*
2³/₁₆in (5.56cm)
Vienna, Austria
Fritz Bermann Co.
Marks/Description:
"FBW" in circle
Ca. 1935
Value: $165

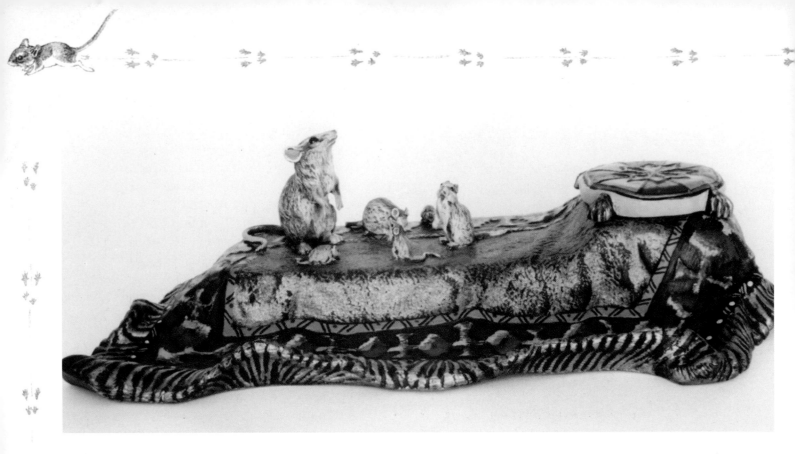

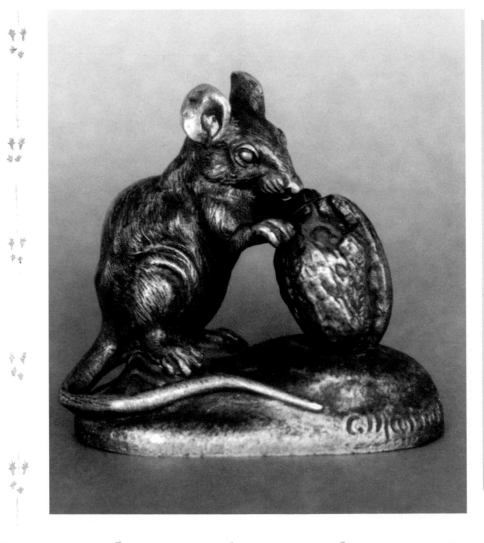

Above:
Vienna Bronze Inkstand with Six Mice
5¾in (14.61cm)
Austria
Marks/Description: hand painted; made in Austria (appears to be Fritz Bermann sculpture)

Ca. 1935
Value: $1,250

Left:
Bronze Mouse with Walnut
3½in (8.89cm)
France
Marks/Description: signed C. Masson. Clovis Edmond Masson was born in Paris in 1838. He studied under Santiago and Barye and exhibited from 1867 to 1881. He won an honorable mention in 1890. He was an animalier. C. E. Masson died in 1890.
Ca. 1890
Value: $450

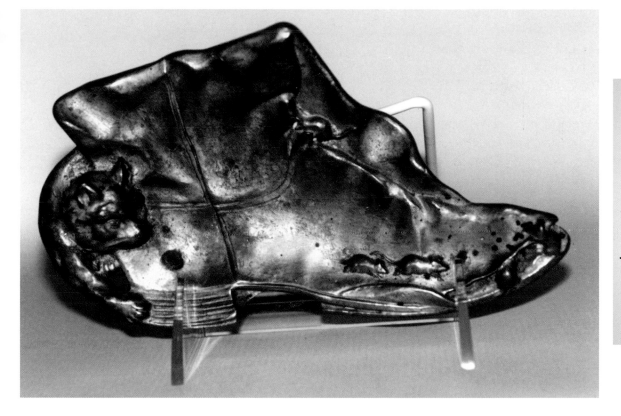

*Bronze Shoe
Ashtray with
Cat and Two
Mice*
7in (17.78cm)
New York, NY
Marks/
Description:
J. B. McCoy
and Son,
New York
   Ca. 1920
Value: $170

*Bronze Mouse on
Stamp Box on
Books; Mouse on
Hinged Book Cover*
2¾in (6.99cm)
Marks/Description:
The name
"Anouctil" is
engraved, and on
one book cover the
word "Album" is
engraved
   Ca. 1900
Value: $325

Mouse and Cat Dancing
2½in (6.35cm)
Vienna, Austria
Fritz Bermann Co.
Marks/Description: "FBW" in circle
Ca. 1935
Value: $225

Bronze Round Container with Mouse Finial
4³/8in (11.11cm)
Austria
Marks/Description: A. Pohl; flower engraving around base. Adolf Joseph Pohl was born in Vienna in 1872. He studied and worked as a decorative sculptor. In 1900, he won an honorable mention at the Exposition Universelle. Listed in *The Dictionary of Sculptors In Bronze*.
Ca. 1900
Value: $250

Three Attached Bronze Mice
1½in (3.81cm)
Geschozt, Austria
Marks/Description: Geschozt, Austria
Ca. 1900
Value: $135

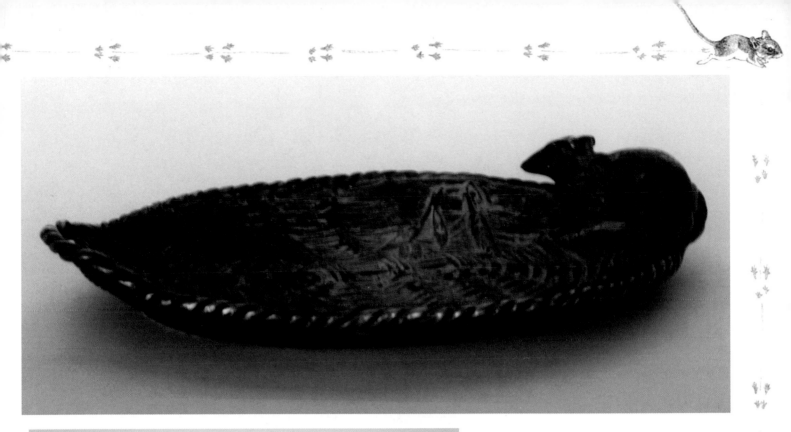

Bronze Ashtray with Mouse
5¹/8in (13.02cm)
USA
Marks/Description: AGPA; Gorham Co. founders; patina finish
Ca. 1920
Value: $135

Mouse with Cheese
1in (2.54cm)
Austria
Fritz Bermann Co.
Marks/Description: none
Ca. 1930
Value: $195

Vienna Bronze Hand-Painted Mice
Mr.: 2in (5.08cm) and Ms.: 1¾in (4.45cm)
Vienna, Austria
Fritz Bermann Co.
Marks/Description: "FBW" in circle
Mr. Mouse with Rifle
Ms. Mouse in Red Dress
Ms. Mouse in Pink Dress with Mop
Mr. Mouse Smoking Pipe
Mr. Mouse with Rake
Ms. Mouse in Blue Dress with Basket
Ms. Mouse in Red Dress with Umbrella
Mr. Mouse Eating Sausage
Ms. Mouse in Green Dress with Basket
Ca. 1930
Value: $225 each (set of nine pieces $2,000)

Bronze Mouse and Camel
Vienna, Austria
Marks/Description: "FBW" in circle; Fritz
Bermann Co.

Ca. 1935
Value: $310

Bronze Mouse Bookends
6in (15.24cm)
Korea
Marks/Description: San Pacific,
San Francisco
Ca. 1993
Value: $85

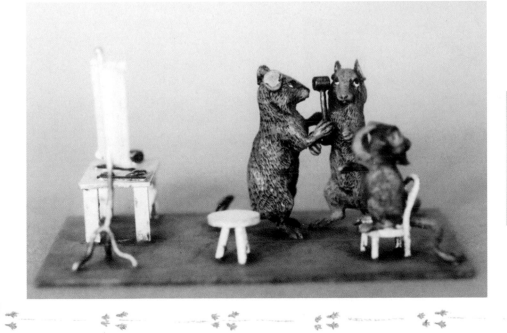

Optometry Office
Austria
Marks/Description:
Austria PR
Ca. 1975
Value: $850

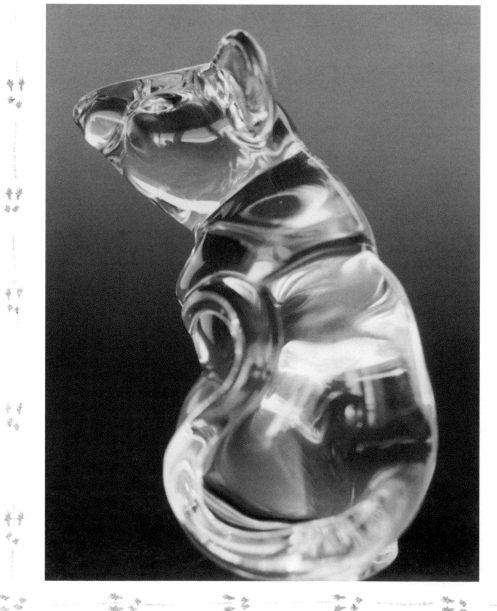

Above:
Crystal Decorative Mouse
5in (12.7cm)
USA
Marks/Description: none
Ca. 1980
Value: $65

Left:
Crystal Mouse
3in (7.62cm)
France
Marks/Description: diamond signature; Daum, France. The Daum signature in various forms was usually signed on the base. The mark became Daum-Nancy with the cross of Lorraine on the base around 1910. France was added after 1919.
Ca. 1980
Value: $190

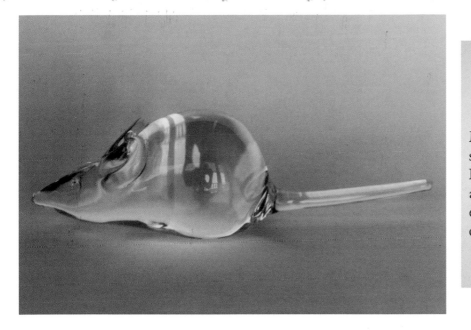

Steuben Crystal Field Mouse
#81314
6in (15.24cm)
Corning, NY
Marks/Description: diamond point
signature "Steuben". Designed by
Lloyd Atkins: Mouse fashioned from
a single gather of molten glass with
eyes indented, ears and tail drawn
out.
Ca. 1978
Value: $225

Multi-Colored Glass Mouse on
Crystal Ashtray
France
Marks/Description: Daum, France
Ca. 1993
Value: $295

Amber Crystal Mouse
3in (7.62cm)
Sweden
Marks/Description: none
Ca. 1985
Value: $12

Steuben Crystal Woodland Mouse
USA
Jane Osborn Smith
Marks/Description: modeled in profile with its
nose resting on its paws and its tail curled to one
side; diamond point signature "Steuben" #8639J
Ca. 1992
Value: $135

Smoky Crystal Lalique Mouse
2³/₁₆in (5.56cm)
France
Marks/Description: R. Lalique, France
Ca. 1932
Value: $775

Crystal Lalique Mouse
2³/₁₆in (5.56cm)
France
Marks/Description: R. Lalique
Ca. 1932
Value: $775

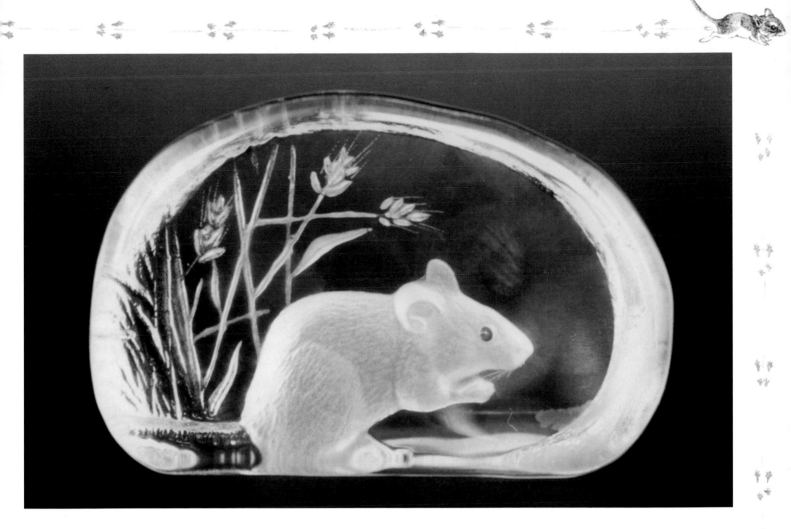

Above:
Full Lead Crystal Mouse
4½in (11.43cm)
Sweden
Marks/Description: Signature Collection; Mats Jonasson; #3262. V. This piece has been made at the Maleras factory founded in Sweden. Mats Jonasson's new technique for sandblasting has gained him world-wide recognition.
Ca. 1985
Value: $145

Mouse in Crystal Globe
4in (10.16cm)
USA
Marks/Description: Maude and Bob St. Clair
Ca. 1980
Value: $100

Fat Brass and Crystal Mouse
Marks/Description: none
Ca. 2001
Value: $28

Brass and Crystal Mouse
1½in (3.81cm)
Marks/Description: letter "S" stamped on base
Ca. 2001
Value: $20

Pearl/Gold Mouse Pin (top)
¾in (1.91cm)
USA
Ca. 1980
Value: $150

Jade 14K Gold Mouse Pin (bottom)
25mm
Japan
Ca. 1980
Value: $225

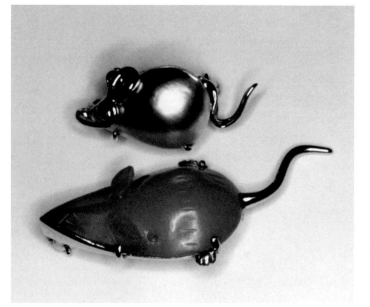

Tennis Player Pin
1½in (3.81cm)
USA
Marks/Description: 18K gold,
platinum, diamonds and pearl
tennis player; sapphire eyes;
letters "EW" & Co.; #750; crown
Ca. 1992
Value: $3,300

Enamel on Gold Mouse Pin
1⅝in (4.13cm)
USA
Marks/Description: #750; ruby eye
Ca. 1985
Value: $650

Enamel Over Gold
Ice-Skater Pin
1⁵/₈in (4.13cm)
USA
Marks/Description: 14K gold;
diamond eyes; blue enamel
Ca. 1990
Value: $650

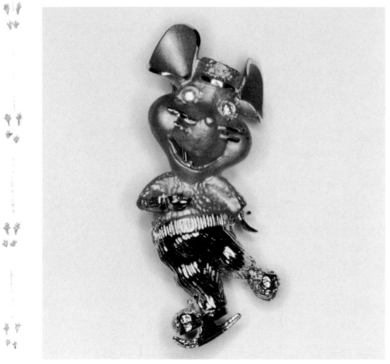

Gold Ice-Skater Pin
USA
Marks/Description: 14K gold;
diamond eyes
Ca. 1990
Value: $650

*Gold Bar with
Snail and Mouse
Looking at Diamond*
1¾in (4.45cm)
USA
Marks/Description:
18K gold
Ca. 1920
Value: $650

Costume Jewelry, Mouse Pin
1½in (3.81cm)
USA
Marks/Description: exceptionally well made with faux diamonds, faux pearls and faux sapphires.

Ca. 1990
Value: $65

Large 14K Gold Ring
USA
Marks/Description: "K" & "V"; ruby eyes; #373

Ca. 1969
Value: $1,350

Small 14K Gold Ring
USA
Marks/Description: 14K gold
Ca. 1970
Value: $350

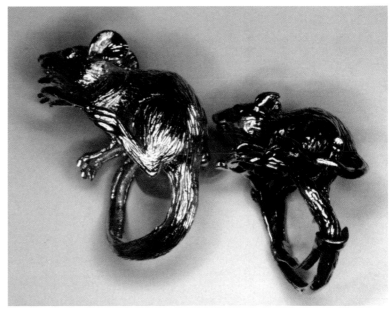

Florentine Gold Mouse
with Ruby Eyes
1³/8in (3.49cm)
USA
Marks/Description: mouse holding diamond; 18Kgold
Ca. 1980
Value: $850

Gold Mouse
1¹/₈in (2.86cm) (not including tail)
USA
Marks/Description: 18K gold;
#6991414; letters "SBI"; sapphire
eye

Ca. 1993
Value: $850

Diamond Mouse with Ruby eyes
USA
Marks/Description: 14K gold (mounting for pin or pendant)
Ca. 1925
Value: $3,400

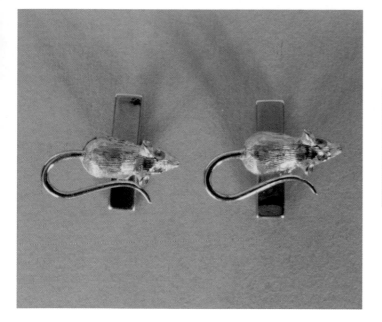

Key Bar Pin with Gold Mouse and Diamond
2³/₁₆in (5.56cm)
USA
Marks/Description: two ruby eyes; diamond;
18K gold

Ca. 1970
Value: $650

Cowboy Mouse Pin
1³/₈in (3.49cm)
USA
Marks/Description: 18Kgold; ruby eye; one cut
diamond

Ca. 1993
Value: $975

Mouse Pin
USA
Van Cleef & Arpels
Marks/Description: ruby eye;
diamonds surrounding eye;
turquoise body; 14K gold
Ca. 1985
Value: $825

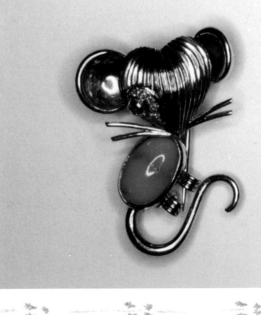

Mice and Egg Ivory Japanese Carving
3¹/₈in (7.94cm)
Japan
Marks/Description: Japanese signature
Ca. 1910
Value: $550

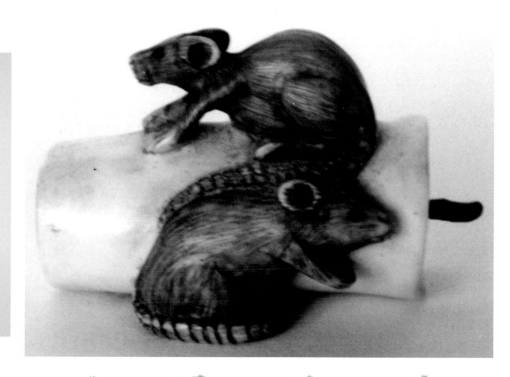

Two Hand-Carved Ivory Mice on a Candle
1³/₈in (3.49cm)
Japan
Marks/Description: Japanese signature. The miniature sculpture, which was developed in Japan over 300 years ago, serves both functional and aesthetic purposes. This Japanese art form is called a Netsuke.
Ca. 1890
Value: $450

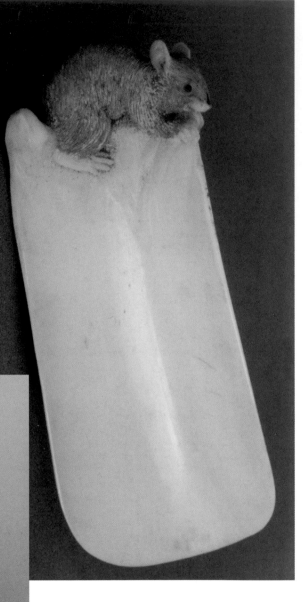

Hand-Carved Bone Shoehorn
with Mouse
4½in (11.43cm)
Marks/Description: none
Ca. 1890
Value: $140

Japanese Ivory Carving
1¾in (4.45cm)
Japan
Marks/Description: Japanese
signature on base
Ca. 1940
Value: $230

Ivory Carving
2in (5.08cm)
Japan
Marks/Description: none
Ca. 1925
Value: $250

Ivory Carving
Japan
Marks/Description: hand-carved ivory
accordion and mouse
Ca. 1900
Value: $110

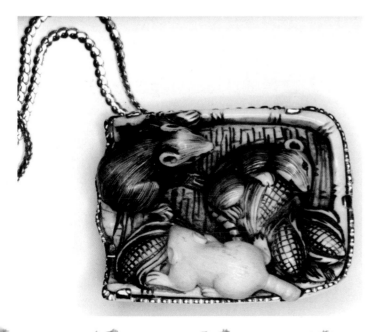

Three Mice Eating Corn
Japan
Marks/Description: ivory sculpture with 14K gold
mounting for pin or necklace and gold chain
Ca. 1950
Value: $850

Pewter Box with Two Mice
2½in (6.35cm)
Marks/Description: none
Ca. 1915
Value: $55

Pewter Tuba Player
1⅝in (4.13cm)
USA
Marks/Description: Hudson Pewter; #166;
letters "HW"

Ca. 1985
Value: $35

Lovers
1⅝in (4.13cm)
USA
Marks/Description: pewter; WF Hudson
Ca. 1980
Value: $70

Pewter Northern Grasshopper Mouse
1½in (3.81cm)
USA
Habitat: Western U.S.
Marks/Description: C-A-79 GM 1500
Ca. 1980
Value: $80

Pewter California Pocket Mouse
1³/8in (3.49cm)
USA
Habitat: California
Marks/Description: C-A-79 CM 1500
Ca. 1980
Value: $80

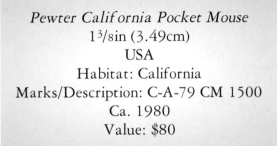

Pewter White-Footed Mouse
1⁵/8in (4.13cm)
USA
Habitat: N, S, E, and Central U.S.
Marks/Description: C-A-79 WM 1500
Ca. 1980
Value: $80

Hand-Painted Pied Piper
and Friends
2½in (6.35cm)
England
Marks/Description: pewter
Ca. 1980
Value: $145

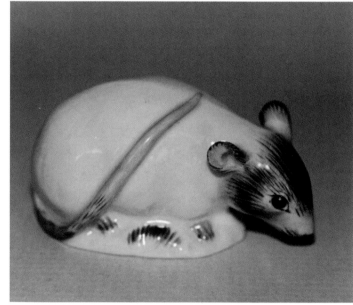

Porcelain White Mouse
Meissen, Germany
Marks/Description: brown eyes and head; #29A
and official crossed swords trademark on base
Ca. 1960
Value: $550

Porcelain Mouse
1³⁄8in (3.49cm)
Bavaria
Marks/Description: Rosenthal, Bavaria;
crown and roses
Ca. 1935
Value: $145

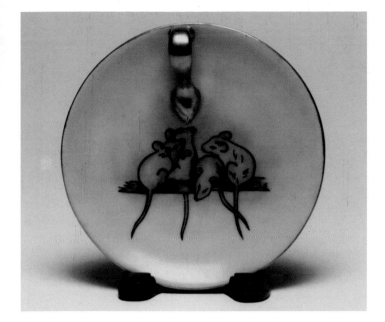

Four Mice on Porcelain Tray with Handle
Bavaria
Marks/Description: F. Harrington, Bavaria
Value: $55

Chester – The Meadow Series
Japan
Fitz & Floyd, Inc.
Marks/Description: "Chester" by Ralph
Watehouse; O.K. Collectibles; Limited Edition,
1983; 61/5000, FF40
Ca. 1983
Value: $250

Mouse Plate
10¼in (26.04cm)
USA
Marks/Description: The American Wildlife
Heritage; John J. Audubon Plate #1290/2500;
from the Audubon original of 1845. Copyright
1977; Volair Limited
Ca. 1977
Value: $160

Right:
Porcelain Mouse Cup
Japan
Marks/Description:
none
Ca. 1995
Value: $12

Below:
*Two White Porcelain
Mice with Shell*
10in (25.4cm)
Eichwald, Bohemia
(Dubi, Czechoslovakia)
Ball Black Co.
Marks/Description:
crown above letter "E"
Ca. 1930
Value: $230

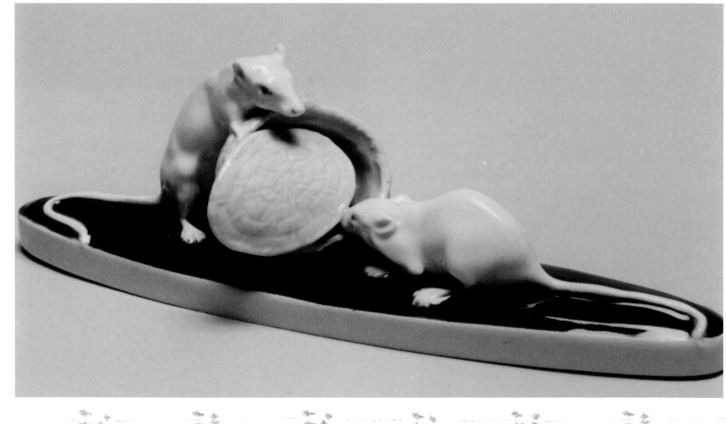

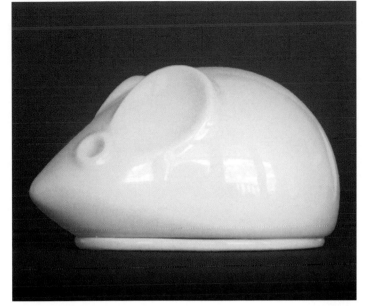

White Porcelain Mouse Container
1¾in (4.45cm)
Arzberg, Bavaria (West Germany)
Company in business since 1975; taken over and
now known as Hutschenreuther.
Marks/Description: Arzberg, Germany
Ca. 1975
Value: $60

*Two White Porcelain Mice on
Ear of Corn*
3¹/8in (7.94cm)
England
Marks/Description: #2017; signature G. L.
Ca. 1950
Value: $150

*Porcelain White Mouse
on Ear of Corn*
5¹/8in (13.02cm)
Copenhagen, Denmark
Marks/Description: pink
eyes; Denmark; #572;
Royal Copenhagen
Ca. 1960
Value: $125

Two Porcelain Gray and White Mice
1¾in (4.45cm)
Denmark
Royal Copenhagen Porcelain Factory
Marks/Description: crown and
insignia #521
Ca. 1960
Value: $125

*Porcelain Mouse Eating
Lobster*
3½in (8.89cm)
Germany
Marks/Description:
T. Hinge; E. Mueller;
#08826 and official seal.
European Art Pottery.
Ca. 1975
Value: $950

Porcelain Mouse Box
2in (5.08cm)
France
Limoges
Marks/Description: Peintmain;
Limoges, France; Sinclair 1999
Ca. 1999
Value: $125

Gray and White Porcelain Mouse
3³/₈in (8.57cm)
Meissen, Germany
Marks/Description: Meissen with crossed
swords
Ca. 1960
Value: $225

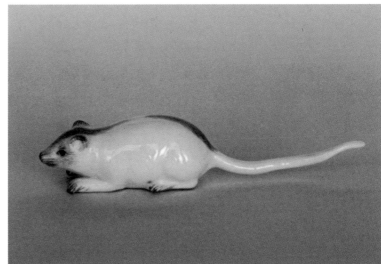

Porcelain Thimble
1in (2.54cm)
England
Royal Doulton
Marks/Description: "At last the sun began to sink
behind the far woods. It was time to go home.";
inside: Royal Doulton: Spring. Jill Barklem, 1983
Ca. 1983
Value: $35

White Porcelain Mouse with
Pink Eyes on Green Shoe
Germany
Marks/Description: Made in Germany;
souvenir of New York
Ca. 1935
Value: $90

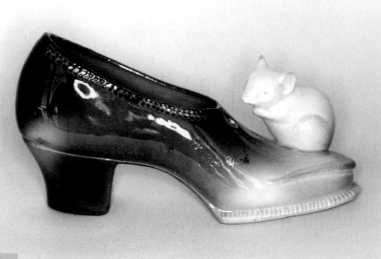

Porcelain Gray Mouse
2½in (6.35cm)
Selb, West Germany
Marks/Description: Hutchenreuther, Selb; lion in
oval with initials "LHS"; Kunstadter; inverted 3
Ca. 1935
Value: $275

Porcelain Bisque White Mouse
2½in (6.35cm)
West Germany
Marks/Description: Kaiser; W. Germany:
Trademark: Incised #595
Ca. 1950
Value: $125

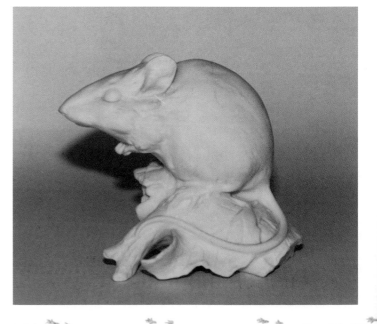

Cookie Jar
USA
Marks/Description:
Los Angeles, CA.;
Laurie Gates
Ca. 1995
Value: $65

Mouse on Gold Ball
3in (7.62cm)
Germany
Rosenthal
Marks/Description: porcelain
white mouse, underglaze
green; crown and two flowers
Ca. 1925
Value: $150

*Two White Porcelain
Mice on Blue Shoe*
2in (5.08cm)
Germany
Marks/Description:
#3748; Pfeffer
Porzellan; official seal
Ca. 1960
Value: $145

*Limoges White Porcelain Mouse
with Nut*
2³/8in (6.03cm)
France
Marks/Description: letters "GDA";
France (under glaze)
Ca. 1920
Value: $250

Mouse Box
1½in (3.81cm)
England
Crummles & Rochard
Marks/Description: mouse
covered with autumn leaves;
hand-painted; English
enamel; inside picture of
an acorn
Ca. 2001
Value: $130

Porcelain Mouse
³/8in (0.95cm)
USA
Marks/Description: USA
Ca. 1975
Value: $35

Porcelain Mouse
1½in (3.81cm)
England
Marks/Description: "I am a magical mouse" on
the base; picture of a mouse coming out of a
man's sleeve

Ca. 2000
Value: $120

Porcelain White Mouse
2in (5.08cm)
Austria
Vienna Porcelain Factory
Marks/Description: official seal; Made in Austria;
stamped #1711

Ca. 1950
Value: $145

Porcelain Herend Mice
Hungary
Herend Porcelain Co.
Marks/Description: These mice are manufactured in 8 colors. Each is surface numbered and also imprinted. Around the shield the words Herend and Hungary are printed, Hand-painted.
Ca. 1985
Value: $190 each

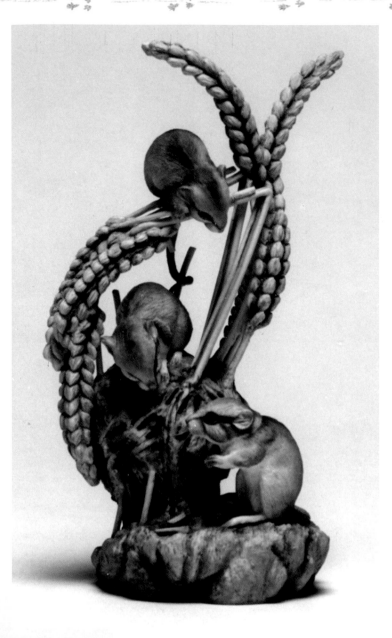

Mouse on Banana
3½in (8.89cm)
Scotland
Border Finer Arts
Company
Marks/Description: '79;
Stock #225057; Schmidt
Sculptured Porcelain; artist
A. Wahl
Ca. 1979
Value: $135

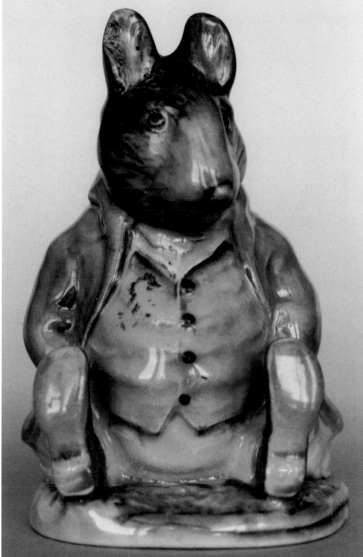

Porcelain Mouse by Beatrix Potter
3¼in (8.26cm)
Beswick, England
Marks/Description: F. Warne & Co. Copyrite, Ltd. Beatrix Potter's Samuel Whiskers; the marks on the bottom of each piece is a sign of a genuine Beswick figure. Beatrix Potter was born in 1866. Her great skill depicted animals essentially naturalistic and lifelike.
Ca. 1985
Value: $65

Mouse Box
1½in (3.81cm)
England
Crummles & Rochard
Marks/Description: hand-painted English enamel; inside: two green leaves
Ca. 2001
Value: $130

Mouse Box
1³/₁₆in (3.02cm)
England
Crummles & Co.
Marks/Description: inside cover has farm scene;
base marked "Made in England" by Crummles &
Co.

Ca. 1985
Value: $190

Porcelain-Covered Container with Mouse Finial
5¹/₈in (13.02cm)
Czechoslavakia
Marks/Description: Schlaggenwald H/C 1792;
1928 stamped on base; 12071 stamped under lid.
Ca. 1928
Value: $180

Crummles Mouse Box
2in (5.08cm)
England
Marks/Description: English hand-crafted:
roses trademark; F & W & Co., 1992
Ca. 1992
Value: $65

Woodland Bone China Cup
4½in (11.43cm)
England
Marks/Description: Roy Kirkham;
English Fine Bone China: Woodland
Ca. 1985
Value: $54

Porcelain Field Mouse
4in (10.16cm)
England
Marks/Description: Designer Barbara
Mitchell, Fine Bone China: Made in
England: Wren Giftware
Ca. 1985
Value: $35

Tea Cup
USA
Marks/Description: company name
incised; illegible
Ca. 1996
Value: $40

Porcelain Brown and White Mouse on Cork
3½in (8.89cm)
Staffordshire, England
Marks/Description: on cork "Moet Chandon",
Imperial 959; on base Bill, MSC, 1987, "C"
Ca. 1987
Value: $75

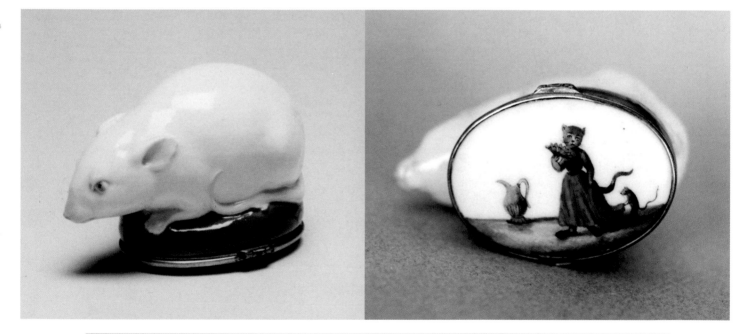

English Porcelain Mouse Box
2³/16in (5.56cm)
England
Marks/Description: inside hinged cover; four mice pulling cat in cradle; cat and mouse
on bottom. George III period Staffordshire porcelain box in the shape of a mouse.
Ca. 1820
Value: $2,400

Three Mice on Tray
6⁷/₈in (17.46cm)
Limoges, France
Marks/Description: tray shaped like a leaf;
"D" and "C"; France
Ca. 1895
Value: $150

Seven Mice on a Porcelain Plate
8⁵/₈in (21.91cm)
Austria
Marks/Description: M. Lashbrook, Austria
over crown over two axes
Ca. 1930
Value: $65

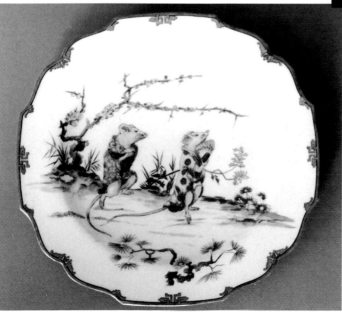

Plate
9in (10.16cm)
Paris
Ch. Pillivuyt Co.
Marks/Description: Medailles d'or, 1867-1878
Ca. 1940
Value: $65

Porcelain Deer Mouse Decorated
3½in (8.89cm)
USA
Edward Marshall Boehm Studio
Marks/Description: signature in feather and horse head #400-89

Ca. 1980
Value: $175

Porcelain Blue Vase with Two White Mice
5³/₈in (13.65cm)
Germany
Marks/Description: under glaze is blue printed "Copenhagen" stamped Germany; crossed swords with letters "RC" stamped below crown
Ca. 1950
Value: $125

Porcelain Gray Mouse
1¾in (4.45cm)
Copenhagen, Denmark
Bing & Grondahl
Marks/Description: "B&G" Castle #1801 incised
Ca. 1960
Value: $150

Flambé Mouse on Cube
2½in (6.35cm)
England
Marks/Description: Royal Doulton Flambé
#1104

Ca. 1912
Value: $900

Flambé Mouse with Nut
England
Marks/Description: Royal Doulton Flambé
#1164

Ca. 1912
Value: $900

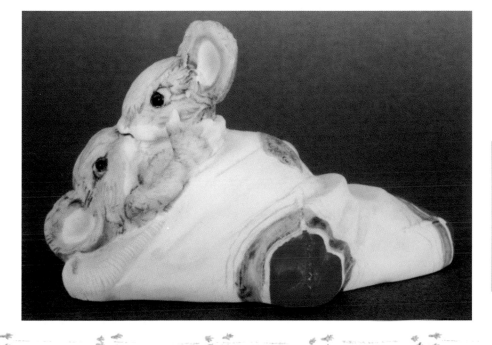

Porcelain Lovers
1in (2.54cm)
Italy
Marks/Description: Artiface:
Decorato a Mano, Otanta
Ca. 1980
Value: $15

Right:
Brown Mouse
2in (5.08cm)
Germany
Hutschenreuther
Marks/Description: lion facing left;
Germany; #104
Ca. 1935
Value: $125

Below:
Swiss Cheese Shape Cigarette Box
Japan
Marks/Description: "Treasure Box"
exclusively made for Recco Collection
Ca. 2000
Value: $30

Porcelain Mouse on Tree Branch
1³/₈in (3.49cm)
London, England
Marks/Description: Asprey Co., Albany, England
Ca. 1980
Value: $225

Porcelain Gray Mouse
2½in (6.35cm)
Germany, Selb, Bavaria, W. Germany
Marks/Description: Hutchenreuther, Selb; lion
with initials "L.H.S." Germany, Kunstadter;
inverted #3

Ca. 1935
Value: $275

Porcelain White Mouse
3½in (8.89cm)
USA
Edward Marshall Boehm Studio
Marks/Description: Signature in feather
and horse head #400-89
Ca. 1979
Value: $175

Above:
Cookie Container
12¼in (31.12cm)
USA
Marks/Description: number 450
Ca. 1990
Value: $24

Left:
Colorful Porcelain Figure
2¾in (6.99cm)
Hungary
Marks/Description: two official seals; below one seal the words "Hungry and Hollohaza"; 1777 is incised. The word "chase" is imprinted
Ca. 1998
Value: $195

Porcelain Teapot Mouse with Blue Eyes
6¼in (15.88cm)
England
Marks/Description: Sunshine Ceramic,
1986-1989. Decorated with blue flowers.
Ca. 1987
Value: $90

Teapot
Otagiri, Japan
Woodbrier Cove
Marks/Description:
Tracy S. Flickinger; incised
Ca. 1995
Value: $35

Coffee Pot
England
Brambly Hedge
Marks/Description: Royal Doulton; Jill Barklem
signature; Royal Doulton bone china; crown and
lion facing left
Ca. 1990
Value: $110

Right:
Snuff Bottle
Japan
Marks/Description: none
Ca. 1950
Value: $100

Below:
Chocolate Lovers
England
Connoisseur
Marks/Description: #80 of
edition of 100; fine bone china
Ca. 1990
Value: $850

The Beatrix Potter Series
4½in (11.43cm)
England
John Beswick Studio Sculptures
Marks/Description: hand
decorated; porcelain; F.
Warne; from the *Tales of
Johnny Townmouse*
Ca. 1984
Value: $35

Three Blind Mice
Thailand
The Greenwich Workshop Collection
Marks/Description: signature and trademark; a work of art in porcelain; artist James Christenson
Ca. 1985
Value: $290 (set)

Tiffany Clock
2½in (6.35cm)
England
Designed by Tiffany & Co.
Bilston & Battersea
Marks/Description: Halcyon Days
Enamels (figure of mouse on each
side of clock)
Ca. 1985
Value: $150

Hand-Painted Porcelain Mice
3in (7.62cm)
England
Marks/Description: Royal
Stradford, Staffordshire, England
(Mr. Mouse) Painted by hand by
"S" & "M"
(Ms. Mouse) Painted by hand by
"B" & "V"

Ca. 2001
Value: $220 (Pair)

Earthenware Teapot
England
The Beswick Co. (Now a part of Royal
Doulton; was established in
Staffordshire, England in 1936.)
Marks/Description: pink ears, nose and tail;
Made in Beswick, England
Ca. 1980
Value: $100

Mouse in Clay Log
USA
Marks/Description: illegible signature
Ca. 1970
Value: $65

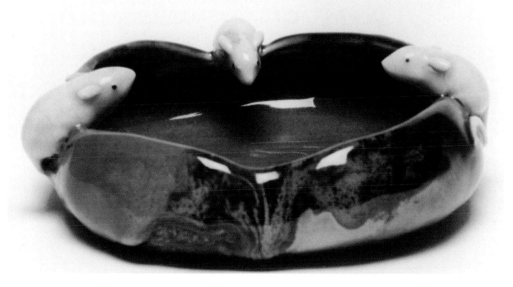

Banko Blue-Green Dish
with Three Mice
5in (12.7cm)
Korea
Marks/Description: Korean
signature on bottom
Ca. 1920
Value: $150

Clay Mouse with Flowers
2³/8in (6.03cm)
Marks/Description: none
Ca. 1980
Value: $12

Green Rosemead Plate
6¹/8in (15.56cm)
USA
Marks/Description: Rosemead
Ca. 1920
Value: $55

Blue and White Posy
Holder with Flowers Cat
Looking Up at Mouse
5¹⁄8in (13.02cm)
Marks/Description: #3208
incised
Ca. 1925
Value: $195

Above:
*Mouse on
Terra-Cotta Gourd*
3¾in (9.53cm)
Japan
Marks/Description: none
Ca. 1900
Value: $145

Great Big Mouse
This two foot tall friend of mine is the first one to greet you when you enter my foyer.

Terra-Cotta Mouse Family in Wheat Field
Marks/Description: none
Ca. 1985
Value: $100

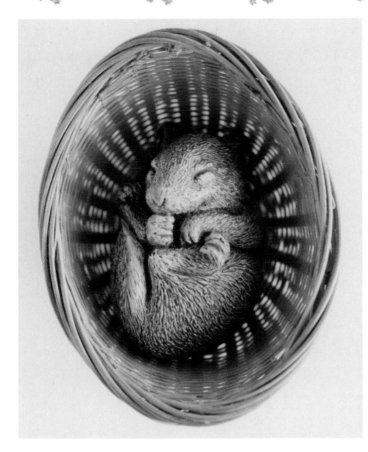

Left:
Sanicast Mouse in Basket
2³/₈in (6.03cm)
USA
Marks/Description: Sanicast Co.
Ca. 1985
Value: $35

Below:
Two Brown Mice with Eggshell
2³/₄in (6.99cm)
England
Marks/Description:
Toquay, Vatcombe
Ca. 1920
Value: $225

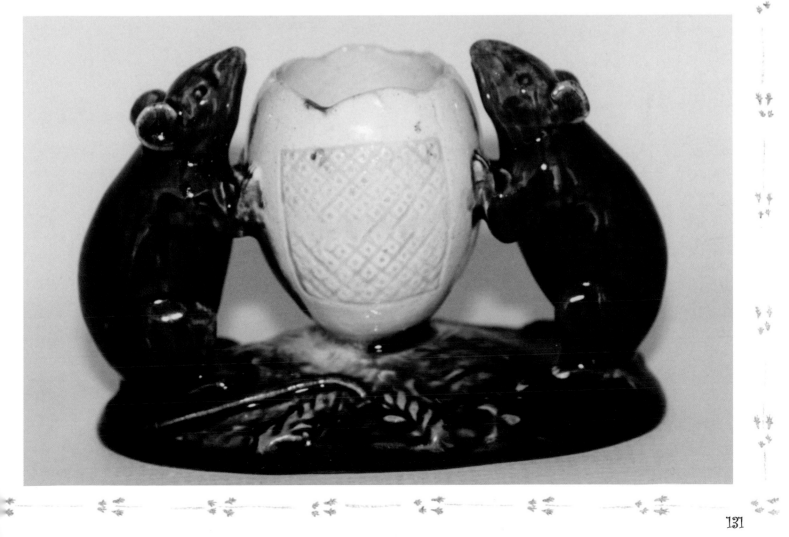

Potpourri Container
Japan
Marks/Description: Made
in Japan
Ca. 1990
Value: $7

Child's Nursery Rhyme Cup
USA
Marks/Description: Dickory,
Dickory, Dock
Ca. 1970
Value: $35

Pied Piper Toby Mug (small)
England
Marks/Description: Pied Piper D6514; Doulton &
Co.; limited; official seal; ©1960. Doulton made
his first pottery and porcelain in Burslem,
England after 1882. The name "Royal Doulton"
appeared on their figures after 1902. The jugs
are modeled and painted by hand, often made in
more than one size.

Ca. 1953
Value: $145

Pied Piper Toby Mug (medium)
England
Doulton & Co.
Marks/Description: Pied Piper D6462; Copr.
1953; Doulton & Co.; limited; official seal;
©1957

Ca. 1953
Value: $145

Pied Piper Toby Mug (large)
England
Doulton & Co.
Marks/Description: Pied Piper D60403; Copr.
1953; Doulton & Co.; limited; official seal
Ca. 1953
Value: $175

Right:
Humorous Mouse
3³/₈in (8.57cm)
USA
Beasties of the Kingdom
Marks/Description: John Raya;
hand-painted
Ca. 1985
Value: $20

Below:
Mice/Mushrooms
England
Chesterton Collectables
Marks/Description: #4912
Ca. 1995
Value: $125

Wall Decoration
6⁵/₈in (17.46cm)
Japan
Marks/Description: several Japanese characters;
can hold flowers or other decorative pieces
Ca. 1985
Value: $45

Bike Buddies
9½in (24.13cm)
China
Marks/Description: no manufacturers name
or remarks except, "Don't Use Near Food"
Ca. 2002
Value: $110

*Three Mice on Silver
Match Holder and Striker*
3in (7.62cm)
Marks/Description: none
Ca. 1900
Value: $175

*Tiffany Sterling Silver
Table Setting*
England
Sugar 6¹/8in (15.56);
salt & pepper 2³/8in (6.03cm)
Marks/Description: sterling;
lion facing left; letter "T";
face of cat; letter "K"; head
of sugar mouse hinges back
to reveal gold-filled spoon.
Ca. 1908
Value: $475

Sterling Silver Wine Taster with Mouse in Bowl

Gorham Sterling Silver

Marks/Description: Trademark registered Dec. 19, 1899. The lion facing right, with the G inscription, was employed by Gorham since 1865. Sometime after this date, a capital letter was used to denote its date of manufacture.

If you look closely, you can see a little silver mouse sitting on the bottom of the bowl. The original purpose of the wine taster can be traced back to the time of the middle ages, to see if the beverage contained any poison. Another use was to aid the vintner in sampling the wine he expected to purchase. This is the reason it was necessary to have a small shallow bowl. This was so the clarity of the wine color could be observed. These silver tasters appeared as early as the fourteenth century. It is almost impossible to find an original today.

After discussions with silver experts, the consensus of opinion regarding the mouse on the bottom of the bowl is as follows: It seems that the vintners discovered that the vineyards that produced the greatest abundance of grapes and the best tasting wine came from those fields that were considerably more infested with field mice than other areas. The conclusion was that the mice aided the growth and development of these specific vines by eating the dead leaves and branches. This selective natural pruning made for better grape orchards and eventually greater volume and better tasting wine. As a memorial, a sterling ladle was manufactured with a mouse sitting on the bottom of the bowl. . .very rare. As a matter of fact, this is the only one that I have ever seen.

Ca. 1890
Value: $1,200

Silver-Plated Napkin Ring Holder
USA
Marks/Description: triple-plate (manufacturer
and trademark illegible)
Ca. 1900
Value: $325

Sterling Silver Cartier Mouse
New York, NY
Marks/Description: sterling, Cartier
Ca. 1975
Value: $190

Sterling Silver Mouse
1¹/₈in (2.86cm)
USA
Marks/Description: diamond
eyes; incised letters "IP";
#88
Ca. 1930
Value: $125

Right:
Cat and Mouse Match Safe
2⅝in (6.67cm)
Austria
Marks/Description: none. We have limited our mouse collection to non-violent pieces. Nowhere will you find a mouse in trouble with other animals or traps. Some figures are plain and simple, some serious, some comical. This figure is the exception to the rule. Finding it to be such a rare item, we decide to include it in the collection. . .besides, if you look very carefully, you can see that the mouse has a smile on his face.

Ca. 1900
Value: $325

Below:
Silver-Plated Mouse on Bronze Purse
5¾in (14.61cm)
France
Marks/Description: L. Carvin; Patrouillfa; Edit Paris: on bottom, Depose\Ca. 1900; Louis Albert Carvin was born in Paris. He studied under Fremiet and Gardet and exhibited from 1894 to 1933. His early work was largely animalier in concept.
Ca.1900
Value: $1,150

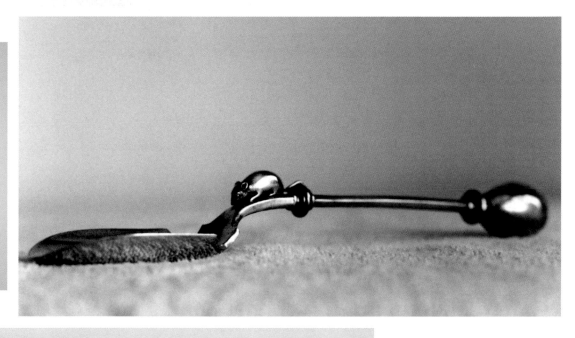

Sterling Baby Spoon and Fork
3¼in (8.26cm)
USA
Marks/Description: international silver
and trademark (mice figures in handles)
Ca. 1950
Value: $135

*Sterling Silver
Mouse on Sterling
Silver Cheese Scoop*
8½in (21.59cm)
USA
Bigelow,
Kennard & Co.
Marks/Description:
sterling, Bigelow,
Kennard & Co.
Ca. 1895
Value: $474

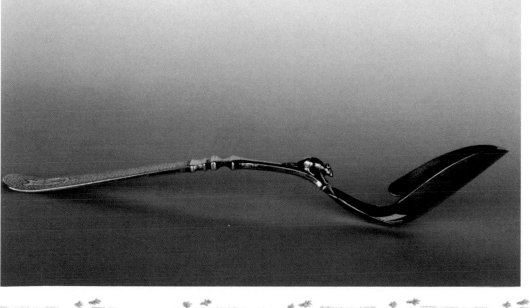

Cheese Scoop
New York, NY
Ball Black Co.
Marks/Description:
Ball Black Co. Sterling;
Initials "RF"; Sterling
silver mouse on handle
Ca. 1880
Value: $350

Right:
Sterling Silver
Mouse
1½in (3.81cm)
USA
Marks/Description: incised #925
Ca. 1930
Value: $60

Below:
Sterling Silver Mouse on Marble
5in (12.7cm)
England
Marks/Description: standing lion and
other symbols indicating city of origin,
maker, and year of manufacture;
stamped "Filled"
Ca. 1965
Value: $625

Sterling Silver Mouse
USA
Marks/Description: none. Purchased because
it was sterling silver and well made.
Ca. 1990
Value: $90

Sterling Silver Tie Clasp
2in (5.08cm)
USA
Marks/Description: sterling; engraving of one
large and one small mouse; letters "WS"
Ca. 1925
Value: $90

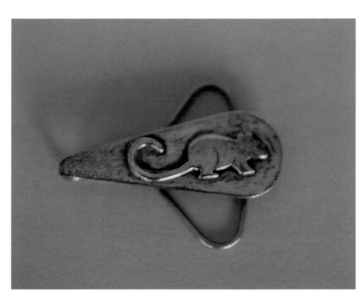

Sterling Silver Bowl
4¹/₈in (10.48cm)
USA
Starr & Marcus
Marks/Description: sterling; #750-A;
lion and anchor; four mice for legs
Ca. 1900
Value: $300

Scratchboard Drawing
4½in x 6⅞in (11.43cm x 17.46cm)
USA
Marks/Description: "It's a Small World" by Peg McInnis
Ca. 1992
Value: $200

Metal Tape Measure
2³/₁₆in (5.56cm)
(without tail)
USA
Marks/Description: red
eyes; Deponirt; turning
tail winds up tape
Ca. 1920
Value: $125

Rhodochrosite Mice
USA
Sweet Home Mine, CO
Marks/Description: two mice and ear of corn carved from
this piece of rhodochrosite mineral (stalactite)
Ca. 2002
Value: $625

Hand-Carved Wooden
Mouse Eating Acorns
on Wooden Leaf
8⁷/₈in (22.54cm)
USA
Marks/Description:
none
Ca. 1925
Value: $175

Composition Mouse on Apple
4¾in (12.07cm)
England
Marks/Description: artist signature illegible
Ca. 1974
Value: $85

Jade Mouse
1in (2.54cm)
Japan
Marks/Description: none
Ca. 1980
Value: $160

Plastic Red Mouse on Stand
1⅞in (4.76cm)
Japan
Marks/Description: none
Ca. 1950
Value: $35

*Leather Purse with Mouse and
Frog Stamping*
3½in (8.89cm)
Chicago, IL
Marks/Description: compliments
of Limacher Buffet
Ca. 1925
Value: $95

Doric Clock
USA
Made by Elias Ingraham
Marks/Description: The Doric
Mantel Clock was patented in
1861. This clock was built upon
two circles. One formed the dial,
and the other formed the base
with the pendulum. It has a
reverse glass painting with a cat
chasing four mice. Manufactured
in 1890.

Personal Mouse Cap

I surgically removed the mouse head from the rest of the body, glued it on the cap and wear it when attending antique shows. This makes it easier for the dealers to recognize me.

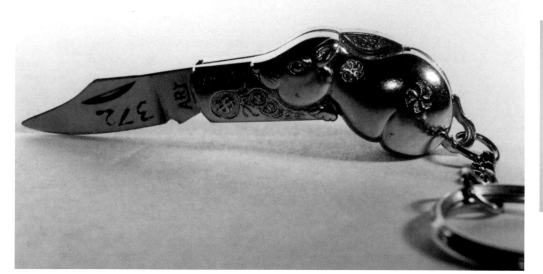

Penknife
Japan
Marks/Description: name "Art" engraved on handle, dollar sign and Japanese letters also engraved.
Ca. 1985
Value: $1

Hand-Painted Metal Mouse and Biscuit
2½in (6.35cm)
Marks/Description: crown and initials "B H" and word "biscuit"
Ca. 1930
Value: $120

Glass Mouse Cologne Bottle
6⅝in (16.83cm)
Germany
Marks/Description: #0175;
Gesgesch
Ca. 1920
Value: $50

Alabaster Mice
Large Mouse 3½in (8.89cm); Small mouse 3in (7.62cm)
Italy
Marks/Description: one of the smaller mice has red eyes
Ca. 1975
Value: Large $45; Small $25

Eschen Scrimshaw
2½in (6.35cm)
USA
Marks/Description: signature
Ca. 1970
Value: priceless

Domestic Mice Lithograph
6⁷/₈in (17.46cm)
Marks/Description: signed, "Lizars, S.C."

Ca. 1880
Value: $300

Acrylic Mouse Earrings
USA
Marks/Description: none
Ca. 1992
Value: $35

Japanese Katani Vase
Marks/Description: Japanese
signature on base
Ca. 1895
Value: $350

Mouse in Moccasin
4½in (11.43cm)
England
Chesterton Collectables
Marks/Description:
Mouse/Moccasin#4913
Ca. 1985
Value: $105

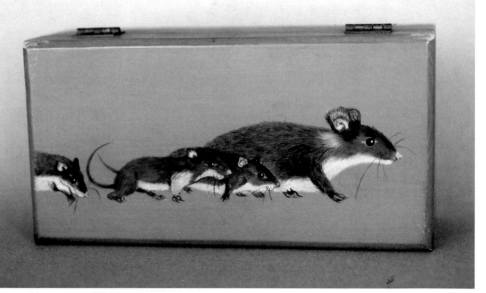

Hand-Painted Cigarette Box
Stonington, CT
Clare R. Bray
Marks/Description: mother
and four baby mice
Ca. 1981
Value: $35

Three White Mice
Marks/Description: none
Ca. 1900
Value: $10

Iron Mouse
3¹/₈in (7.94cm)
USA
Marks/Description: none (made with
nuts, bolts, nails, and washers)
Ca. 2002
Value: $5

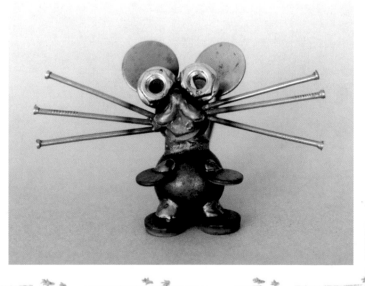

Book Published 1890
Three Old Friends
USA
Marcus Ward & Company
Ca. 1890
Value: $90

Wax Mouse
5³/₈in (13.65cm)
USA
Marks/Description: ©1983
Ca. 1983
Value: $12

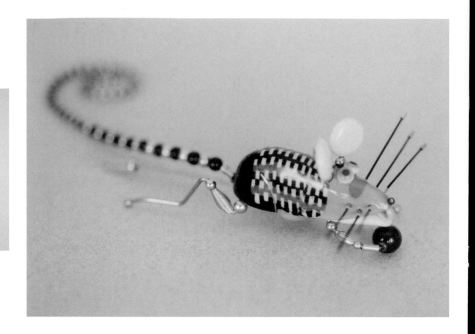

Contemporary Mouse
Manhasset, N.Y.
Cynthia Chuang, 10+ Gallery
Marks/Description: mice; different
colors; different shapes
Ca. 2000
Value: $140

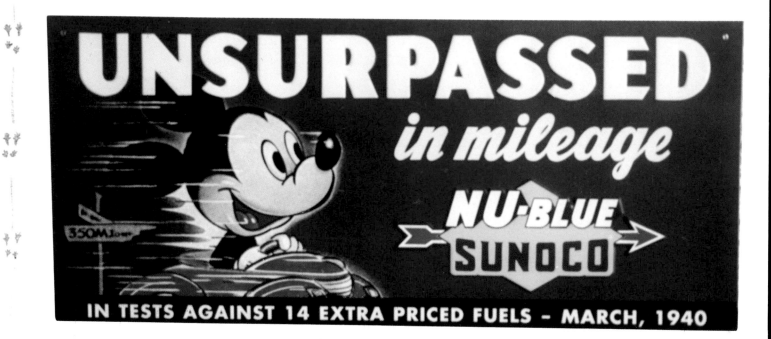

Sunoco Gasoline Commercial
Evidently this is a copy. Not a true antique
Marks/Description: none; measures 8 inches x 18 inches
Value: $20

Mouse Soap Dispenser
USA
Marks/Description: none
Ca. 2002
Value: $12

Bookends and/or
Doorstop
Japan
Marks/Description:
several Japanese symbols;
manufactured in three
sizes; made of iron
Ca. 2000
Value: $5 - $10

Mouse (named Finky)
2³/₁₆in (5.56cm)
England
Harmony Kingdom
Marks/Description: "Finky"; Harmony Kingdom;
Copyright 2000

Ca. 2000
Value: $50

White Avon Perfume Bottle
3¹/₈in (7.94cm)
Marks/Description: #6; Avon
Ca. 1975
Value: $15

Mouse Box
3¹/₈in (7.94cm)
England
Harmony Kingdom
Marks/Description: inside is figure of a trap; six
symbols on base; "The Mouse That Roared" also
on base; Copyright 1997; initials; Jr. holding baby
mouse in hands

Ca. 1997
Value: $80

Glossary

ALABASTER—A compact fine textured gypsum, which is usually white and translucent.

ALLOY—A combination of metals fused together.

BACCARAT—A French glass company that has produced fine-quality glassware since 1765.

BATTERSEA—Painted enamel, the product of a company set up about 1753 by Stephen Janssen in Battersea, England.

BELLEEK—Irish china, thin and light in body, highly translucent, with a cream-like surface in ivory tone.

BESWICK—A manufacturer of collectible figurines. Part of Doulton Company.

BILSTON—Town in Staffordshire where decorated enamelware was produced during the 18th century.

BING & GRONDAHL—Makers of porcelain in Copenhagen, Denmark.

BISQUE or BISQUIT—A fired ware which has neither glaze nor enamel applied to it.

BLOWN GLASS—Glass that is blown rather than pressed or molded.

BODY—The clay, which comprises the earthenware.

BOEHM—Company that manufactures porcelain figurines of animals, flowers, and birds in Trenton, New Jersey.

BONE CHINA—After 1880 the mixture of bone ash with kaolin and feldspar. The mainstay of the English porcelain industry from 1820.

BRASS—A metal alloy of copper and zinc.

BRONZE—An alloy of copper and tin. First metal used by man.

CELLULOID—An early, very flammable plastic.

CERAMICS—Clay items fired at high temperature.

CHASING—A process in which hammers are used to create decorative details without removing the metal.

CHINA—Originally referred to all wares coming from China. Today this generic term means products that are fired at a high temperature (also known as porcelain).

CLAY—A general term used for the material used to make ceramic items.

COALPORT—Porcelain factory founded about 1796 by John Rose.

COPENHAGEN—City where soft-paste porcelain factory was founded about 1759. The hard-paste factory was established in 1774.

CYBIS—Excellent quality figurines produced in Trenton, New Jersey.

DAUM NANCY—Glassworks producing cameo-style glass in the art nouveau style.

DELFT—A tin glazed pottery decorated with blue on a white background.

DERBY—Company in England producing pottery. Royal Crown Derby is modern name.

DIAMOND POINT ENGRAVING—Designs scratched in glass by the point of a diamond.

DOULTON—Ceramic company established in 1815 in Lambeth, England. After 1902 known as Royal Doulton.

DRESDEN—The porcelain produced in Dresden, Germany, during the 1700s and 1800s. Crossed swords means that the item was made in the Meissen factory.

EARTHENWARE—The oldest ceramic substance. Most often glazed and fired.

EDITION—The number of items made with the same number and decoration.

EMBOSSED—Raised ornamentation.

ENAMEL—The opaque semi-transparent colored substance that is used in coating metals and porcelain; later fired.

ETCHED—Cutting into the surface to produce decorative designs with acid.

FAIENCE—Pale red earthenware covered with a tin glaze.

FINIAL—An ornament figure that tops a vase or cup cover.

FISCHER, MORITZ—Founder of a pottery factory in Herend, Hungary.

FLAMBE—A vivid red-and-purple ceramic glaze.

GALLE—Founder of a ceramic firm in Nancy, France, in 1874.

GATHER—The blob of glass at the end of a blow-pipe before the glass has been blown.

GLASS—A brittle substance made by fusing potash and sand.

GLAZE—The coating applied to earthenware and porcelain when in the bisque stage—before firing.

GOEBEL—Ceramic manufacturer who started the Hummelwork Porcelain company in 1871.

HALLMARKS—The term established to prevent fraud and to set standards of purity.

HARD-PASTE PORCELAIN—Technical term for porcelain made according to the Chinese formula.

HIMOTASHI—The holes made in netsuke so they can be strung on a cord.

HUTSCHENREUTHER—A porcelain manufacturer in Selb, Germany, established in 1814.

INCISED—A design that is cut or engraved into the surface of an object.

IVORY—True ivory carvings are made from elephant tusks.

JADE—Material used to make jewelry, figurines, and, accessories.

KAOLIN—Chinese name for clay.

LALIQUE—A french glassmaker who produced glass usually with a frosted finish.

LEAD CRYSTAL—The brilliance in glass due to the lead oxide used in manufacturing.

LENOX—Ceramic company started by Walter Lenox in 1906.

LIMOGES—A French city famous for its porcelain manufacturers.

LIMITED EDITION—An item produced only in certain quantities or only during a certain period of time.

LLADRO—Quality porcelain figurines manufactured in Spain since 1951.

MAJOLICA—Lead or tin-glazed brightly decorated pottery. The name derived from the island of Majorica.

MAKERS MARK—Identification marks made with either a metal stamp or impressed, or painted or printed-on each piece.

MALACHITE—A dark green mineral used for stoneware; a source for copper.

MEISSEN—Quality porcelain. Started in Germany in 1710.

MINTON—A ceramic factory formed by Thomas Minton in 1796 in Staffordshire, England.

NETSUKE—A Japanese carved figure with holes used to secure the cord on which a person carried his personal belongings.

NORITAKE—A Japanese ceramic company established in 1904.

NYMPHENBURG—Hard-paste porcelain made at a factory in Bavaria in 1747.

OPALESCENT—Glass with an iridescent look.

PASTE—The mixture of ingredients from which porcelain is made.

PATINA—Originally the greenish surface film on copper-containing metals found in the course of time by oxidation.

PETUNSTE—A substance mixed with clay used in producing hard-paste porcelain.

PEWTER—An alloy of 85% tin with other metals. Pewter darkens and dulls easily but rarely oxidizes.

PONTIL MARK—A mark or scar usually left on hand-blown glass objects when the rod is removed.

ROSEMEADE—A North Dakota porcelain company noted for their animal figures from 1949 until 1961.

ROSENTHAL—A ceramics company established in Bavaria about 1879 and still in operation.

ROYAL COPENHAGEN—A pottery founded in Denmark in 1772 and still in operation.

ROYAL CROWN DERBY—This company has been in operation since 1890 making fine quality porcelain.

SABINO—Art glass made in France in colored, frosted and opalescent colors.

SAND CASTING—A special method for casting metal.

SCRATCHBOARD DRAWING—A fine art technique resembling wood engraving.

SCRIMSHAW—A procedure in which fine, detailed etchings are made primarily on elephant tusks.

SEVRES—Porcelain factory founded at Vincennes in 1738. Produces finest French porcelain.

SILVERPLATE—Manufacturing process in which pure silver is electroplated onto a base metal.

STAFFORDSHIRE-An area in England where many pottery factories operated in the early 1700s.

STERLING—Sterling is 925 fine. The standard for early English silver.

STEUBEN—Glass company, part of Corning since 1918, manufacturing extraordinarily fine art glass.

TERRA-COTTA—A reddish earthenware, glazed or unglazed, fired clay.

TIFFANY GLASS—Hand-made iridescent glass called favrile.

TOBY JUG—Drinking mug in the shape of a person.

WHITE METAL—Any of several white alloys usually tin, copper, and zinc (imitating silver).

WEDGWOOD—Major pottery factory established in England in 1730.

ZSOLNAY—Pottery factory established in Hungary about 1862.

Bibliography

Anderson, Hans Christian. *Classic Fairy-Tales*. Morris Plains, New Jersey: The Unicorn Publishing House, Inc.,1993

Bartlett, John. *Bartlett's Familiar Quotations*. Boston, Massachusetts: Little, Brown & Co., 1980.

Berling, K. *Meisen China*. New York, N.Y. (An Illustrated History): Dover Publications, 1972.

Bingham, Don & Joan. *Tuttle Dictionary Of Antiques & Collectibles*: Rutland, Vt: Charles E. Tuttle Co. 1992.

Bond, Harold L. *An Encyclopedia of Antiques*. New York, N.Y: Tudor Publishing Co., 1945.

Browning, Robert. *The Pied Piper of Hamlin*. Avenel, New Jersey: Outlet Book Co., 1993.

Buten, David. *Wedgwood*. (Guide To Marks And Dating), Merion, Pennsylvania: Buten Museum of Wedgwood, 1976

Cosentino, Frank J. *Edward Marshall Boehm*. Chicago, Illinois: R. R. Donnelley & Sons., 1970.

Cosentino, Frank J. *The Porcelain Art of Edward Marshall Boehm*. New York, N.Y: Frederick Fell, Inc., 1960

Dale, Jean. *The Charlton Standard Catalog of Royal Doulton Figurines*. Birmingham, Michigan: W.K. Cross, 1993.

Eichenberg, Fritz. *Endangered Species*. Mills, Maryland: Stemmer House Publishers, Inc., 1979.

Eliot, Marc. *Walt Disney, Hollywood's Dark Prince*. New York, N.Y: Carol Publishing Group, 1993. *Funk & Wagnall's Standard Dictionary of Folklore (Mythology And Legend)*. Maria Leach, Editor. New York: Funk & Wagnall's, 1972, 1950,1949.

Godden, Geoffrey A. *British Pottery and Porcelain*. London, England: Frederick Warne & Co. Ldt. 1973

Hollis, Richard & Sibley, Brian. *The Disney Studio Story*. New York, New York: Crown Publishers, Inc, 1988

Hornby, George. Editor, *Poems for Children and Other People*.-New York, New York: Crown Publishing Co. Inc., 1975.

Hrabalek, Ernst. *Wiener Bronzen* Vienna, Austria: Verlag Laterna Magica. 1991.

Jacobs, Joseph and Joslyn. *The Child's Story Book*. New York, New York: EP. Dutton Publishers, 1987.

Jenkins, Dorothy H. *A Fortune in the Junk Pile*.- New York, New York: Crown Publishers, Inc.,1963.

Kovel, Ralph M & Terry H. *Dictionary of Marks; Pottery and Porcelain*. New York, New York: Crown Publishers, Inc.,1953.

Kovel, Ralph & Rovel, Terry. *Kovel's Antiques & Collectibles Price List, 1990*.- New York, New York: Crown Publishers, Inc., 1990.

Kovel, Ralph & Terry. *Know Your Collectibles*.New York, New York: Crown Publishers, Inc., 1981.

Kovel, Ralph & Terry. *Kovel's .New Dictionary of Marks*. New York, New York: Crown Publishers, Inc., 1986.

Lang, Gordon.,Consultant, *Miller's Antique Check List*. New York, New York: Penguin Group, 1992.

Lukins, Jocelyn. *Doulton Flambe Animals*. London, England: M.P.E. 26, Chapel Lane, Horrabridge, Yelverton, Devon.

Mackay, James. *The Dictionary of Sculptors in Bronze* Woodbridge, Suffolk:Antique Collectors Club, 1977.

Madigan, Mary Jean. *Steuben Glass*. New York, New York: Harry, N. Abrams, 1982.

Malting, Leonard. *The Disney Films*. New York, New York: Crown Publishers, 1984.

Marks, Mariann Katz *Majolica Pottery*. Paducah, Kentucky: Collector Books, 1983.

Moore, Clement C. *The Night Before Christmas*. Racine, Wisconsin: Western Publishing Co., 1982.

Oxford University Press. *Dictionary of Quotations*. London, England: Geoffrey Cumberlege Publisher To The University), Second Edition, 1952.

Palley, Reese. *The Porcelain Art of Edward Marshall Boehm*.- New York, New York: Harry N. Abrams, Inc.,1976.

Perrault, Charles. *Cinderella*. Mahwah, New Jersey: Troll Associates, 1979.

Potter, Beatrix. *The Beatrix Potter Treasury*. New York, New York: Viking Penguin Group, 1988

Plaut, James S. *Steuben Glass, A Monograph*. New York, New York: Dover Publications, Inc., 1972.

Rainwater, Dorothy T. *Encyclopedia of American Silver Manufacturers*. New York, New York: Crown Publishers,

Rice, D.E. *English Porcelain Animals of the 19th Century*.- London, England: Antique Collectors Club, 1989.

Schwartzman, Paulette. *European And American Art Pottery*.- Paducah, Kentucky: Collector Books, 1978.

Shertzer, Hazel. *Aesop's Fables*. U.S.A: Book Sales Inc., 1980

Singleton, Esther. *The Collecting of Antiques*. New York, New York: The Macmillan Co., 1947.

Webster's Encyclopedic Unabridged Dictionary Of The English Language., New York: Dilithium Press, 1989.

About the Author

Dr. Albert H. Eschen, born and raised in Brooklyn, N.Y., now lives in Florida, with his wife Flori. Eschen completed his undergraduate work at Brooklyn College, and then enlisted in the U.S. Air Force where he served as an instrument flying instructor. After his discharge, he returned to school, and four years later received his Doctorate in Optometry from Northern Illinois College of Optometry. Dr. Eschen is a past president of the Brooklyn Optometric Society, and former director of the Brownsville Boys Club Eye Clinic. He was the first pediatric optometrist appointed to the New York City Department of Health. Named Alumnus of the year in 1993, and received the Presidential Medal of Honor from Illinois College in 1999. Avocations: engraving, writing, antique collecting. Prize possessions are his two sons (Burt, an Optometrist practicing in Brooklyn, New York, and Andrew, an attorney) and two grandchildren, Meryl and Lowell.